A SHEARWATER BOOK

THE RISING SEA

The Morris Island Lighthouse in Charleston, South Carolina, was built in 1876 approximately 1,600 feet (500 m) behind the beach. Today, the lighthouse stands about the same distance out to sea. The shoreline here has retreated about 3,200 feet (1000 m) since the lighthouse was constructed, mainly because of the sand-trapping effect of jetties at the harbor's entrance. Much erosion on sandy shorelines today is due to engineering structures and navigation channel maintenance, but in future decades, sea level rise will become the dominant cause of shoreline retreat.

The
RISING
SEA

Orrin H. Pilkey *and* Rob Young

ISLANDPRESS / Shearwater Books
Washington | Covelo | London

A Shearwater Book
Published by Island Press

Copyright © 2009 Orrin H. Pilkey and Rob Young

All rights reserved under International and Pan-American Copyright
Conventions. No part of this book may be reproduced in any form or by
any means without permission in writing from the publisher: Island Press,
1718 Connecticut Ave., NW, Suite 300, Washington, DC 20009.

SHEARWATER BOOKS is a trademark of The Center for Resource Economics.

Library of Congress Cataloging-in-Publication Data
Pilkey, Orrin H., 1934–
The rising sea / Orrin H. Pilkey and Rob Young.
p. cm.
"A Shearwater Book."
Includes bibliographical references and index.
ISBN-13: 978-1-59726-191-3 (cloth : alk. paper)
ISBN-10: 1-59726-191-2 (cloth : alk. paper)
1. Sea level. 2. Coast changes. I. Young, Rob. II. Title.
GC89.P48 2009   363.34'93—dc22
2009006152

Printed on recycled, acid-free paper ✹

Design by David Bullen

Manufactured in the United States of America

10   9   8   7   6   5   4   3   2   1

Keywords: Sea level rise, West Antarctic ice sheet, Greenland ice sheet,
shoreline erosion, flooding, barrier islands, global warming, climate change,
climate skeptics, coastal hazards, coastal management

To Walter Pilkey
for a lifetime of friendship and inspiration

To David Robert Young
who taught me more than he would ever know

Contents

Preface

During the past 2.5 million years, massive continental ice sheets advanced and retreated many times across planet Earth's northern hemisphere. With each advance, the amount of water bound in ice increased, and the level of the sea dropped. With each retreat of the ice sheets, meltwater was released, and the level of the sea rose. Over this time, ocean level has fluctuated across a range of more than 500 feet (150 m), and shorelines have moved landward or seaward tens of miles as a result. In fact, sea level change has been a constant part of earth history as long as there has been an ocean.

So, why all the fuss about the seemingly small sea level rise today? Simply put, the difference is us. Modern sea level rise is encountering for the first time a densely developed shoreline, putting the ways of life of millions of people at risk.

As the current rate of sea level rise accelerates, it imperils our cities, ports, and resorts that are jammed up against the shore. Most of this massive infrastructure is virtually unmovable, or very difficult to move. Future flooding of some cities such as Miami and Singapore is a certainty with rises as little as 2 feet (0.6 m), as is the mass migration of large numbers of refugees from low-lying delta regions to higher ground. In some cases, plans are afoot to move entire cultures back from the sea, such as the Alaskan Inupiat Eskimos and the atoll dwellers of the Maldives in the Indian Ocean and Tuvalu in the Pacific Ocean. Several experts now believe that some communities in the Mississippi Delta should be moved in their entirety to new and higher sites.

In one sense, the human species has been through this before. There is ample evidence of Native American settlement on what are now submerged continental shelves. The ruins of ancient Alexandria on the Nile Delta and other once important cities lie submerged on the floor of the Mediterranean Sea. Alexandria fell into the sea, not as a result of a gradual sea level rise but because of a catastrophic and instantaneous sinking of the land surface during an earthquake. Ancient migration routes like the one across the Bering Strait between Siberia and Alaska were submerged and effectively closed by the rising sea many thousands of years ago.

The current sea level rise of about ⅛ inch (0.3 cm) per year is not perceptible to the casual observer. And because it's not visible, it doesn't impress. But anyone who frequents the coast can see much evidence of recent sea level change. For example, entire island communities have disappeared from parts of the Chesapeake Bay. On Portsmouth Island on North Carolina's Outer Banks, a cemetery used by early English settlers has become a salt marsh, while the old pipes that are supposed to drain surface water runoff from South Carolina's Charleston Peninsula are now partially blocked at high tides. And in the Marshall Islands, salinization of the soil caused by rising sea levels has halted centuries-old gardening practices. Vegetables once grown in small family plots now are planted in abandoned fifty-gallon oil drums filled with soil. And geologists who study the polar ice sheets of Greenland and Antarctica have been recording changes that have momentous implications.

All indications are that we should be alarmed about the future of sea level rise and should be doing something about it now. We chose to write this book because we believe the public needs to have a clear guide to the critical but basic facts about sea level rise and its implications, in order to make intelligent decisions. The existence of a huge "manufactured-doubt" industry is part of the reason for the relative lack of societal concern about sea level rise. In fact, we were both unaware of the extent of this industry until we started researching this volume. One book written for children even argues that melting of the Antarctic and Greenland ice sheets won't affect sea level rise much. The author is actually off by 200 feet (61 m), which is the amount the sea level will rise if the ice sheets melt!

Between us we have sixty-five years of experience studying marine and coastal processes, and we both have long been involved in the societal debates over eroding shorelines. Rob Young spent his childhood on the Virginia shores of Chesapeake Bay, where he was fascinated by stories of disappearing islands. As a doctoral student at Duke University under Pilkey's supervision, Young studied the response of wetlands to sea level rise along the shores of Pamlico and Albemarle sounds in North Carolina. Wetlands such as salt marshes move inland in response to rising sea level in spurts, he discovered, corresponding either to forest fires or to storms from the sea. Forest fires removed the vegetation that resisted shoreline retreat, and storms overcame that resistance.

Orrin Pilkey grew up in the desert of Washington State, far from the sea. Even though he lived near the West Coast, his first childhood view of a beach was in Atlantic City, New Jersey, where feeding the pigeons left the strongest impression. Pilkey has looked at sea level rise evidence and the mode of island response on many of the world's barrier islands. He has observed Colombian Pacific fishing villages where sea level rise rates are so high (due to sinking land) that homes have been designed to be moved to a new location by a small work crew without the help of any machinery. Both authors have investigated heavily developed coasts from Dubai to Daytona, where sea level rise will have a huge impact on the beaches lined with immovable seawalls and high-rise buildings.

In the text that follows, we hope to make the case that the world is poised on the edge of a cliff (of its own making). We must act now by responding to the challenges of sea level rise in a planned and rational way, taking a long-term view. If we don't start planning now, a huge "natural disaster" is facing us. It comes down to accepting the challenge of the rising sea or ignoring it until it is too late and we drive over the cliff.

The
RISING
SEA

— Chapter 1 —

Living on the Edge

All the rivers run into the sea; yet the sea is not full. . . .
Ecclesiastes 1:7

A RISING SEA is not something that may happen in the future. It is already upon us. Planners turned down construction of a large residential development on the Yorke Peninsula, South Australia, because it would be flooded by rising seas. In England, regulators declared that six small villages on the Norfolk Broads northeast of London will need to be abandoned as sea level rises. To avoid the rising sea, the 580 Inupiat Eskimo inhabitants of Shishmaref, Alaska, will likely be moved to the mainland at a cost of several hundred thousand dollars per resident. On barrier islands along the Pacific coast of Colombia where the sea level is rising with particular rapidity (because the land is also sinking), moving buildings and entire villages to higher ground is already a routine matter. In rural Cape Town, South Africa, a "blue line" may be established seaward of which nothing can be built because it would lie within an expected flood zone from sea level rise. Plans to abandon many Pacific atolls are now on the drawing board because they will soon be flooded by the expanding oceans. And in South Carolina, retreat from the shoreline in response to the sea level rise is now official state policy.

Still, despite strong evidence of global warming and attendant sea level rise, many communities, governments, and developers continue to ignore the inevitability of a continued rise in sea level and the corresponding increase in shoreline erosion. Singapore continues to fill in its bays to create more low-elevation land for development. In a stunning act of developmental hubris, the government of Dubai has constructed spectacular, palm-shaped artificial islands along the Persian Gulf providing space for hundreds of homes, all at low elevation and immediately susceptible to even modest sea level rise. In the United States, the State of Florida seems content to spend billions of dollars in a losing battle to hold the shoreline in place with artificial beaches, breakwaters, and seawalls while high-rise beachfront construction continues apace. And state and federal officials continue to insist that most Mississippi Delta communities can be maintained in their present location despite recent rapid sea level rise augmented by subsidence (sinking) of the land.

Global warming is changing many things: the extent of ice on the surface of the Arctic Ocean, the extent of mountain glaciers, patterns of rainfall and drought around the world, and routes of ocean currents. As the oceans warm, wide swaths of coral reefs, responsible for much of the diversity of marine life, may be degraded as human activities prevent their natural expansion to the north or to the south away from the equator. Shoreline-hugging and biologically important salt marshes and their warm-water equivalent, mangroves, already seriously reduced by the activities of humans, will further degrade as sea level rises. The distribution and the migration pathways of land mammals, birds, and insects will change, and some species will disappear entirely. Mosquitoes will appear in the high Arctic.

Of all the ongoing and expected changes from global warming, however, the increase in the volume of the oceans and accompanying rise in the level of the sea will be the most immediate, the most certain, the most widespread, and the most economically visible in its effects.

Substantial sea level change will play a critical role in humankind's

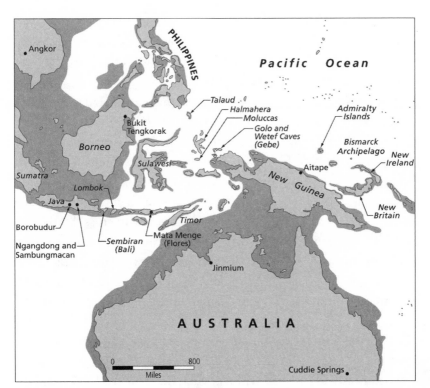

Variation in sea levels in the past has been an important factor controlling the migration paths of humans and animals. The dark areas show several land bridges as they stood about 40,000 years ago, such as the one between New Guinea and Australia across which the Australian Aborigines walked. A number of other land bridges formed when sea level was low as a result of glaciation, including the bridge across the Bering Sea between Asia and North America.

future just as it sometimes has in the past, when it even became the subject of myth. Plato, for example, suggested that nine thousand years before his time an ancient civilization had existed on an island called Atlantis, only to disappear somehow beneath the waves. Some writers today cling to a belief that the Bahama Banks hold the answer to the island's disappearance, that the long, narrow bands of underwater limestone there (actually cemented beach sand) are remnants of either the mythical city's roadways or its building foundations.

Others of the Atlantis faithful believe that a cataclysmic event such as a volcanic eruption destroyed the island, perhaps the same eruption that destroyed the Aegean island of Thera about 1500 BC. Still others believe Plato made up the whole thing.

In any case, it has been established that during a period forty thousand to sixty thousand years ago in which sea level was considerably lower than it is today, Aborigines walked across an exposed land bridge between New Guinea and Australia, and some of these early Australians walked on from Australia to Tasmania. The first Americans may also have taken advantage of a more recent comparatively low sea level, perhaps eleven thousand years ago, to cross the Bering Land Bridge from Asia to North America. Stone Age people at the same time must have crossed back and forth between the British Isles and Europe. Eventually, as we know, the sea rose and covered these ancient access routes, changing the face of the earth and the lives of its people.

Many societies are candidates to be the first in our time to suffer catastrophic impacts from impending sea level changes. Eventually, every nation with a coast will feel the effects of sea level rise, but in this chapter, we tell the stories of a few that are most immediately and tragically vulnerable. It's a cruel irony that many of these societies, for the most part non-industrial, have played almost no role in the global warming that lies behind so much of current sea level rise; in that sense, they are truly innocent victims of the industrialized world.

Arctic Islands: Abandoning a Sinking Ship

The Arctic seaside villages of Alaska are made up mostly of clusters of government-issue small homes. Houses there are tightly closed and heavily insulated with vestibules between outer and inner doors to prevent the extreme Arctic cold from penetrating the house when someone opens the outer door. These houses stand in sharp contrast to the thatched-roof dwellings perpetually open to the ocean breezes on the atolls of the South Pacific that are also in danger from sea level rise.

Satellite measurements show that the level of the sea is rising in the Arctic Ocean, but that's only part of the problem facing these high-latitude shoreline dwellers. In earlier times, Arctic islands were completely protected from the ocean by ice for most of the year. Because of warming atmospheric and oceanic temperatures, ice-free conditions—normally only two to three months in extent—have lengthened to four and even five months along the North Slope of Alaska and even longer along the shores of the Chukchi and Bering seas to the south. Longer ice-free periods subject the shoreline day after day to the high waves associated with fall and winter storms.

Summertime melting of permafrost (permanently frozen and therefore solid ground) underlying shoreline bluffs and beaches compounds the problem: melting of the ice effectively removes the cement that made the beach a natural seawall and thus greatly facilitates erosion at the shoreline. Add an ever-rising sea level to this picture and the situation becomes dire indeed.

Two Alaskan shoreline villages that have garnered considerable attention are Kivalina and Shishmaref. Kivalina is an eight-mile-long (13 km) Arctic barrier island northwest of Kotzebue in the Chukchi Sea. This community of four hundred was originally a winter encampment for Inupiat Eskimos but is now inhabited year-round. Besides a rapidly receding shoreline on both sides of the island, the village suffers from serious local river pollution emanating from the Red Dog zinc mine, located up the Kivalina River on the mainland. The villagers have filed a lawsuit against the Environmental Protection Agency (EPA) claiming the agency is not enforcing clean water regulations; ironically, the mine itself is on Native Corporation land.

Shishmaref, also on the shores of the Chukchi Sea, is located just south of the Arctic Circle. There are 580 people in this subsistence village located on 4-mile-long (6 km) Sarichef Island, one of the islands that make up the Shishmaref barrier island chain.

A close look at Shishmaref's options and deliberations can give

An aerial view of the Inupiat Eskimo village of Shishmaref, Alaska. A combination of sea level rise and melting permafrost in the beach are eroding the shoreline, putting survival of this village at risk. Note the protective rock revetment (seawall) in front of the village (as well as a previously constructed rock structure already underwater). The cost of moving this village of 580 people to the mainland could exceed $100 million. (Photo courtesy of the National Oceanic and Atmospheric Administration)

us some perspective on many of the issues that will confront island beachfront communities everywhere in the uncertain future of global change. For Shishmaref, the future is now; and for the hundreds of other beachfront communities on barrier islands from Maine to Texas, and around the world, the future is not far off.

The plight of Shishmaref has attracted global attention. In fact, it has become the poster child of the Arctic global change and sea level rise problem. Since 2002, more than sixty-five media crews from as far away as Sweden and Japan have come to the relatively remote island. Its particular attractiveness to the media may come from a combination of relatively easy access by small plane from Nome or Kotzebue and the welcoming presence of Tony Weyiouanna, community transportation manager and a spokesperson of exceptional skill.

Shishmaref is a village of contrasts. During a February 2007 visit to the village, accompanied by Alaskan geologist Owen Mason, we entered our rented house through a heat-saving vestibule and stepped past a freshly killed white Arctic hare on the floor. A recently skinned and frozen reindeer carcass hung over the balcony on the front of the house. The house had no running water, and the toilet consisted of a large pail in a bathroom. But the 40-inch (100 cm) TV set in front of two comfortable sofas offered fifty channels.

A winter visit to Shishmaref quickly dispels any doubt that it is a subsistence society. A walk through the village in cold weather is a walk through a giant freezer. In the yards of homes, along with abandoned snowmobiles, dogsleds, four-wheelers, and various derelict household items that peek out from beneath the snow, are seal, caribou/reindeer, and rabbit carcasses in various stages of skinning and butchering. Meat and hides are perched on racks, balconies, and banisters to keep them out of reach of the few dogs allowed to roam freely through the village. Most of the dozens of sled dogs, however, offer no threat to the frozen meat. They remain chained to steel posts even in the coldest weather, with their only pastime for days on end consisting of barking at any and all passersby. Between the barking dogs and the roaring snowmobiles, an Eskimo village in this region can be a very noisy place in winter. And it's not much quieter in summer either, when the four-wheelers zip about with multiple passengers and supplies.

Unlike their Inupiat Eskimo brethren on the North Slope of Alaska along the Beaufort Sea, Shishmaref residents do not consume whale meat as part of their normal diet. The nearby continental shelf of the Chukchi Sea is so gentle and flat that whales don't swim close enough to shore to be taken by harpoons from small boats. The whalebones and baleen carved by these villagers for sale to craft stores in Alaska's cities are from the occasional dead whale that washes up on the beaches.

Primary subsistence foods for Shishmaref villagers include fish (salmon, cod, whitefish, trout, and herring), moose, musk oxen, ducks, geese, ptarmigan, walrus, a variety of berries, and various greens. Two small grocery stores provide bananas, oranges, and other supplemental foods. Some dollar bills in the cash registers are worn almost white, a reflection of a society with limited contact with the outside world.

The tide range on Shishmaref's shores is on the order of 1 foot (0.3 m), and surges during extreme storms could raise the water level as much as 8 feet (2.4 m) above the high tide line. Such a storm surge would flood much of the island, which now has lost its protective oceanfront dune ridge.

Shishmaref's home island of Sarichef, known as Kigiktaq Island before the arrival of Russian explorers in the early 1800s, has been occupied continuously on a seasonal basis for at least four centuries and probably much longer than that. Lieutenant Otto Von Kotzebue named the inlet at the north end of the island Shishmaref after a crew member, a name adopted eventually for the village itself and for the lagoon behind the village. Until the beginning of the twentieth century, the Eskimos may have used the island only as a winter camp; during the warmer season, they spread out to other islands and the mainland to hunt and fish. Their descendants do the same today, occupying summer hunting camps that were established hundreds of years ago, although now there is a year-round population in Shishmaref. At the end of the nineteenth century, the harbor at Shishmaref became a drop-off point for supplies to the goldfields to the south, and the village began its year-round existence.

Today the village, by an inhabitant's description in a 2002 newsletter, is comparable to a third world community: "Most families do not have running water and sewer services in their homes. The lack of roads, high costs of fresh foods, inadequate fuel storage for home heating and transportation, exorbitant costs of basic services and the constant anxiety caused by beach erosion is an excessive burden carried by all members of the community."

Add the multifaceted problems associated with global warming to those of living with financial hardship and one is left asking, So what can Shishmaref do?

Alternative #1.
Hang in there, and build, maintain, and reinforce seawalls.
The problem with this approach is obvious—as the year-round sea ice shrinks and summer (with its attendant open water) lengthens, as storms become more frequent, as more permafrost melts each summer, as sea level rises, and as the waves grow larger, the wall must grow larger or community damage will increase. Since the 1950s, when erosion first became a recognized problem, the community has taken many measures to solve the problem. These measures included using oil drums and sandbags to construct seawalls, and when conditions grew desperate, locals threw everything that normally resided in yards into the raging surf, including kitchen sinks, old dogsleds, and junked snowmobiles, in an attempt to reduce storm impacts.

In 1984, a 1,700-foot (520 m) seawall made of gabions—wire baskets filled with stones—was installed. The baskets fared well, but storms excavated the sand behind the wall and the shoreline continued to retreat, albeit at a slower pace. At one point, an unusual wall consisting of 25-pound (11 kg) cement blocks linked by cables to form a mat was laid up against the dune face. The wall was designed to bend a bit when sea ice pushed up against it. Much of this structure failed in less than a year (parts of it began to collapse even while it was under construction).

As geologist Owen Mason sees it, construction of the various seawalls has actually exacerbated Shishmaref's survival problem. All the natural dune and beach protection has gone, and the narrowed and steepened beach in front of the walls enhances the impact of storm waves. By 2007, the shoreline was retreating on either side of the walled portions of the community. At this rate, soon, perhaps in a decade, Shishmaref will be a mini-cape protruding into the sea, a magnet for waves.

By the end of 2006, around $34 million had been spent on Shishmaref seawalls, including a stone wall costing $16 million. Spending $34 million, with no contribution from the community, for what is at best a temporary solution to the erosion problem for a village of fewer than 600 people is mind-boggling. In the contiguous United States, a favorable cost-benefit ratio to erect seawalls for such a small village could never have been achieved since the cost of the seawalls would far exceed the replacement cost of threatened buildings. But a different standard is applied to subsistence villages.

Alternative #2.
Remain on the island, but move houses to new locations
as they become threatened by erosion.
Eighteen of the small houses of double-wide size were moved back from the shoreline after a 1997 storm. However, protective dunes along the island front have largely been lost in recent decades, and no matter where houses are moved on the island, they will be repeatedly threatened by storm flooding. Wherever the community would move on the island, it would very likely remain huddled behind large seawalls.

If either of the first two alternatives had been chosen as a path forward, it would be necessary to construct storm shelters for community members and to have firm island evacuation plans that could be dependably activated at very short notice (and presumably in very harsh weather).

Alternative #3.
Move to Anchorage, Nome, or Kotzebue.
Of course, moving to a larger community would immediately halt the subsistence lifestyle. Community spokesperson Tony Weyiouanna pleads that "we have lost our language and we have lost our dance but we have not lost the subsistence nature of our society. This we must preserve."

The issue is a complex one. The communities of Nome (population 3,540) and Kotzebue (population 3,230) are willing to take Shishmaref citizens, but what will people with extraordinary hunting and fishing expertise do in the "big city"? In the background, unspoken,

looms the issue of alcoholism. Eskimo native peoples have historically had a serious problem with alcohol abuse, which is why most
Eskimo villages, including Shishmaref, are bone dry by choice of the
inhabitants.

Alternative #4.
Move to nearby native villages.
A larger population would likely damage the local subsistence hunting and fishing as the local resources become overwhelmed by the
increase in population. In addition, there may be long-standing rival-
ries between local populations.

Alternative #5. Move the community to the mainland.
A 2003 General Accounting Office report declared that Shishmaref
is one of the top four Alaskan communities most in need of relocation because of the threat of flooding and further beach erosion.
Apparently, the idea of moving the village was seriously considered as
early as 1973 in response to the first of the big storms that struck the
community that year. The idea snowballed, and Shishmaref villagers
cast ballots on the issue in July 2002. By a substantial margin, they
chose to relocate in full recognition that even if all went smoothly,
it would take fifteen to twenty years to complete the move. Surprisingly, young voters in the village were more reluctant than the elders
to endorse the idea. The chosen site, known as West Nantuq, is to the
southwest on the mainland across a 5-mile-wide (8 km) lagoon.

This is now the "winning" option. All movable buildings will
be transported across the frozen lagoon in midwinter. The requi-
site gravel for roads and building foundations in permafrost on the
mainland will be obtained from a nearby volcanic rock outcrop (on
National Park Service land). The U.S. Army Corps of Engineers' plan
includes construction of an extensive rock jetty to form a harbor for
the new community. This seems to be an add-on feature since the
community thrives now without such a harbor.

According to a 2004 study by the Corps of Engineers, all of Shishmaref's options would be expensive. The cost of relocating to the mainland will be $180 million, while the estimated cost of staying on the island ran to $110 million. Moving the villagers to Nome would require an estimated $93 million, while moving them to Kotzebue would cost an estimated $114 million.

The new village on the mainland will consist of 186 homes. Of existing village homes, 137 will be moved across the frozen lagoon at an estimated cost of $300,000 each, and a few new modular homes will be brought in by barge from Seattle and put in place at $300,000 each. Additional expenses involve utilities such as water, sewage, electricity, and heat, all very costly in the Far North. The federal government is expected to pay for everything.

The estimated costs of moving Kivalina (the sister city to the north) are about the same as for Shishmaref. Town officials in Kivalina seem to agree that the Corps of Engineers' estimates of moving costs are too high. Perhaps this is posturing on the village's part to assuage the fears of the Congress over an immense bill for moving tiny villages. But there is justifiable concern, quietly stated in the villages, that the Corps of Engineers has a long-term and well-deserved reputation for doing things in a very costly fashion and a well-known record of coastal engineering incompetence, recently accented by the Hurricane Katrina levee disaster in New Orleans. Yet this is the agency on the forefront of the nation's response to sea level rise, and this is an agency with deep pockets, so it is where people will turn for solutions.

Kivalina is carrying the torch a bit further. In 2008, the village decided to sue nine oil companies, fourteen power companies, and one coal company to pay for the village's move to the mainland. The villagers' reasoning is that since the companies are in significant part responsible for the shoreline erosion problem, they should be held financially responsible. Kivalina officials, in their legal documents, argue that global warming is caused by the CO_2 these companies produce and that global warming is responsible for the sea level rise that torments the villagers.

In other barrier island settings, mostly in the developing world, moving houses and communities has been used as a successful option by threatened villages for years. It can be a very low-cost option in warm climates if properly prepared for. On the remote Pacific barrier islands of Colombia, sea-level-rise rates are locally very high due to tectonic, or mountain-building, forces associated with the nearby Andes mountain chain. In one study of Soldado Island near Buenaventura Bay, the rate of sea level rise was estimated at 10 feet (3 m) per century, most of it in spurts associated with earthquakes. Small villages are routinely moved after earthquakes (often accompanied by small tsunamis) that cause the islands to sink as much as 3 to 4 feet (0.9 to 1.2 m) overnight. Often, the villagers move laterally to nearby higher islands. Houses are built of panels, which can be quickly dismantled and carried away, panel by panel, to a new location.

Buildings in villages along the Niger Delta in West Africa are also designed for rapid removal to a new location in response to sea level rise. There, sea level rise is very rapid due to subsidence probably caused or at least accelerated by extraction of oil and water from beneath the delta. Shoreline erosion rates along the island fronts in this region are usually on the order of 50 feet (15 m) per year. The Niger Delta islands are the widest barrier islands in the world— usually on the order of 3 miles (5 km) in width—and there is dense rain forest land available to occupy as the population retreats from the coast. In Nigeria as well as in Colombia, one of the great attractions of continuing to live near the beach is the breeze that tends to reduce the quantity of malaria-bearing mosquitoes.

Shishmaref's efforts to survive are backed by the deep pockets of a rich country. The communities in Colombia and Nigeria, by contrast, are virtually on their own. It remains to be seen whether the U.S. government will be willing to spend $300,000 to $400,000 per individual to move a remote Arctic village. Living on the edge of a modern nation, the villages are a unique and rare American native culture treasure. Given this fact, combined with the energy of Tony Weyiouanna and others like him who wish to preserve their culture, it is likely that Shishmaref will survive—in a new location—and perhaps provide an example of resilience that the lower 48 coastal villages may someday envy. Moving to the mainland from barrier islands may be a widespread practice in the lower 48 in a few decades, but coastal communities will likely have to pay for this themselves.

The Pacific Atoll Nations: Nowhere to Go

Atolls are the Hollywood-celebrated, spectacular coconut palm tree–lined rings of small coral islands that surround beautiful blue lagoons. Charles Darwin, musing at the rail of HMS *Beagle,* hypothesized (correctly) that Pacific island atolls were rings of coral left behind as a coral-rimmed volcanic peak sank below the waves. But there is a problem in paradise. By most measures, the coral atoll islands (mostly in the South Pacific) should be number one on the list of nations endangered by sea level rise.

The inhabited portions of atolls lie around only 3 feet (0.9 m) above the level of the sea, just like currently threatened Arctic coastal villages and the endangered zones of Bangladesh. Relative to the 15 million people of Bangladesh threatened with rising seas, however, these island nations have tiny populations. They include Kiribati (28,000 population), the Maldives (269,000), the Marshall Islands (58,000), Tokelau (2,000), and Tuvalu (10,000), among others.

What distinguishes their plight from other threatened nations is that the atoll dwellers have no place to go, no higher ground in their possession for escape. They suffer from what University of Sydney geographer John Connell calls the tyranny of distance. The islands that make up these tiny resource-poor nations are at great distances from one another and at an even greater distance from any large landmasses.

And these tiny countries, most with subsistence cultures, have no additional resources to help them respond to sea level rise (with the possible exception of the Maldives and its high-end tourist economy). In this era of globalization and increasing expected self-reliance, the atoll nations are entirely dependent on the often dwindling generos-

Satellite image of Niau atoll, population 136, in French Polynesia. The origin of atolls by upgrowth of coral reefs around a sinking seamount was first suggested by Charles Darwin. These atolls, some of which have populations in the tens of thousands, will soon be evacuated as a result of the rising sea level. Salinization, storm damage, and destruction of vegetation, rather than actual inundation, will drive the inhabitants away. (Copyright © SSTL through ESA. Used with permission.)

ity of surrounding larger nations such as Australia, Japan, New Zealand, Canada, and the United States. One problem hindering future emigration from Pacific atolls is that the potential receiving nations are more and more favoring immigrants with specific skills, which, of course, the islanders don't have. Just like the problem the Inupiats from Shishmaref face, well-honed hunting and fishing skills are of little use in an industrialized society.

Rising seas are not an abstract notion for atoll communities. These nations face the loss of the very physical basis of their national sovereignty and the loss of their culture. All are contemplating a mass movement of their citizenry. Furthest along in this endeavor are Kiribati, whose capital island is Tarawa of World War II fame, Tuvalu, and the Maldives.

Moving citizens off the islands has already begun on Carteret Atoll (population about 1,600 in 2007), which is near New Guinea and is not a sovereign nation. Carteret Atoll is a horseshoe-shaped scattering of tiny islands with a total area of 0.6 square mile (1.6 km²), lying 260 miles (418 km) from Bougainville, the largest of the Solomon Islands. It is possible that sea level is rising faster here than on most Pacific atolls because Carteret Atoll lies near an active tectonic zone. The islands are rapidly losing area due to shore erosion, and every few months they are flooded by storm surges. The periodic saltwater flooding has killed fruit trees, leaving only coconuts and food from the sea for island staples. In the 1990s, a Taiwanese fishing company almost denuded the island's waters of giant clams, taking away what was once an important food for the local people.

During the Bougainville civil war of the 1990s (involving local islanders seeking independence from Papua, New Guinea), refugees from the war crowded onto Carteret Atoll. The extra population pressure on food resources further damaged the islands' food supply. The situation was worsened by the use of dynamite for fishing on nearby reefs. Such a fishing technique, while producing a bountiful harvest on a particular day, destroys the very environment that nurtures the next generation of fish.

In 1989, with the help of Australia, ten families from these disappearing islands were moved to Bougainville. These were among the first sea-level-rise refugees of our time, forced to relocate because of global warming. The refugees ended up being housed next to a prison, some distance away from their beloved sea, forced into farming of unfamiliar plants employing unfamiliar farming methods and eating strange foods. The families became part of a country with dozens of strange languages and a struggling economy. Observers reported that women from the new village often hiked to the shoreline and stood for long periods of time, staring forlornly out to sea. They soon moved back, at least temporarily, to the atoll from Bougainville, not only because of their poor living conditions but also because the local civil war made a bad living situation all the worse.

Now the government of Papua, New Guinea, is seeking suitable sites to resettle the entire population of Carteret Atoll but is having difficulty finding sites because of competing land ownership claims emanating from the civil war. The Carteret residents fear that they will be settled inland again, away from the sea and the only way of life they understand, just as happened to the first ten resettled families. Still, the evacuation can't happen too soon. Recent visitors reported that some of the children on Carteret Atoll are potbellied and have unusual hair color, sure signs of advancing malnutrition.

The most widely heralded (in the Western press) atoll evacuation problem is that of Tuvalu, a nation of eight atolls halfway between Australia and Hawaii. It is a nation of 9.7 square miles (25 km^2) spread over 289,600 square miles (750,000 km^2) of ocean. Saltwater flooding in residential areas takes place with increasing frequency, and that, coupled with the permanent loss of land during storms, has led to an arrangement with New Zealand to take settlers from Tuvalu. The goal is to gradually move the total population of ten thousand individuals.

More than a third of the population of Tuvalu lives on Funafuti, an island with an area of slightly more than 1 square mile (2.6 km^2). The highest point anywhere on Tuvalu is 13 feet (4 m), but most of the habitable land is a mere 3.3 feet (1 m) above sea level. Over past decades, the population grew faster than agriculture and natural marine resources could keep up, creating an increased need for imported food. Tuvalu has no manufacturing industry or tourism (aside from the recent phenomenon of the global warming tourism trade made up primarily of visiting scientists and members of the media). Ironically, the money sent back home by emigrants from Tuvalu to Australia and New Zealand has been a substantial boost to the nation's economy.

There are some "solutions" to the atoll problem, but they are temporary at best. Increasing use of solar energy would reduce use of local wood for cooking and thereby protect the sheltering mangroves of the atolls. Rain catchments could be improved to reduce reliance on increasingly salty groundwater. To avoid the salty soil on Majuro Atoll in the Marshall Islands, for example, vegetables are being grown in oil drums filled with soil. Protective seawalls or dikes in most instances are too expensive to be a serious remedy and, in any event, would be at best a temporary stopgap.

Some residents of Tuvalu cast the situation as a matter of human rights. They argue that industrialized nations are at fault for creating excess carbon dioxide and causing the level of the sea to rise. Thus, the argument goes, the large, rich, mainly Western countries owe the Pacific islanders a decent alternative. Interestingly, Canadian Inupiat Eskimos, like their American counterparts in Kivalina, Alaska, who have already made the move, are contemplating lawsuits jointly with Pacific Islanders against Western carbon dioxide producers.

Sea level rise is clearly responsible for Tuvalu's significant land loss, much of which has occurred since the 1990s. Not unexpectedly, some problems these islanders face are exacerbated by choices they've made. Some of the shoreline erosion must be due to cutting away the mangrove trees at the shoreline for use as firewood. It is also a certainty that the beaches on Carteret Atoll have been subjected to sand mining for centuries and continue to be—a wheelbarrow load here, a bucketful there—to help with construction of nearby homes. Mining of beaches is a form of beach erosion and is a particularly destructive practice in a time of rising sea level. Unfortunately, beaches are mined all over the world, intensifying the impact of sea level rise.

Not all small atoll nations are helpless. The Maldives, located in the Indian Ocean, for example, are made up of twenty-two coral atolls and more than twelve hundred reef islands. This small nation is a case study of the problems created by fluctuations in sea level. Today, rising sea level is beginning to drown them. Male, the capital city (population about 100,000) is now completely surrounded by a massive seawall funded in part by Japan.

The Maldives nation has a flourishing, high-end tourism industry. Former President Maumoon Abdul Gayoom was rightly concerned about the impact climate change will have on his nation's prosperity. In 1992 at the United Nations Earth Summit, he said, "I stand before you as a representative of an endangered people. We are told that as a result of global warming and sea-level rise, my country, the Maldives, may sometime during the next century, disappear from the face of the Earth." At a time when few people had even heard of climate change, it was a dramatic claim.

Today, the claim doesn't sound extreme. Primarily in response to the threat of sea level rise, now ex-President Gayoom increased the elevation of Hulhumale island to serve as a refuge for his people. A new hospital, new schools, new government buildings, and new apartments were all constructed in Hulhumale on ground several feet higher than the rest of the Maldives. A huge dredge sucked up sand from the ocean floor and disgorged it into a shallow lagoon to create the flood-resistant island. That was in 2003. Now, several thousand people live in Hulhumale. Gayoom's goal was to attract at least fifty thousand.

In 2008, Mohamed Nasheed became the first democratically elected president of the Maldives, and as one of his first acts he announced a plan to buy land, perhaps in Sri Lanka or India, on which the nation can reestablish itself.

Other atoll communities are unlikely to have the finances to mimic the Maldives and plot their own relocation, and one has to ask what life will eventually be like as the natural islands disappear and residents are left stranded on an engineered platform. And what will residents do if the living corals that surround the natural islands disappear with the atolls? Will the tourists, who play such an important role in island economy, still come?

Farewell to Venice

The disasters facing Shishmaref and the Tuvalu atolls involve relatively few people and little infrastructure. In the global scheme of things, they represent one endpoint in the sea-level-rise crisis. Venice, Italy, a fifteenth-century city without a future, is another endpoint. It is the location of perhaps the most visible sea-level-rise crisis in the world today.

Venice is in a lagoon protected from open ocean waves by a chain of three barrier islands. During the last century, the sea level rose 10 inches (25 cm) around Venice. The rise is a result of a combination of the global sea level rise and the subsidence caused by the shifting of tectonic plates, as well as the compaction of underlying sediments due to groundwater extraction. The most salient problem in the city is the flooding that occurs as ocean storms and storm surges roar ashore through the barrier island inlets and into the lagoon. Even the smaller, more frequent floods add to the city's problems with gradual degradation of the building foundations as they are periodically immersed in salt water. What does one do as a historically precious and deteriorating city is flooded with ever increasing frequency?

In 1900, Venice's famous St. Mark's Square was flooded seven times. By 1996, the year of the Great Flood (a particularly large storm on November 4), the annual number of floods had risen to ninety-nine. Today, the square is underwater about a third of the time. On December 1, 2008, the highest flood in twenty-two years occurred. It was the fourth highest ever recorded and was caused by high winds from the Adriatic pushing water into the lagoon.

The impact of these storms has been greatly increased by human activities like the dredging of navigation channels and the construction of sediment-diverting jetties. The situation mirrors some of the problems of the Mississippi Delta, as we'll see in chapter 8.

Recent plans call for the construction of giant gates across the inlets through which the storm surges flow. The idea is to close the $3 billion gates just before the storm arrives. Critics argue that as sea level rises and perhaps as storm frequency increases as part of global warming, the gates will be closed for increasing lengths of time, decreasing the circulation and increasing pollution in the already highly polluted lagoon.

The population of Venice dropped from 121,000 in 1996 to 62,000 in 2009 and is still declining. The number of tourists that arrive daily is approaching that of the city's population. At what point should this treasure that is Venice be abandoned?

For Shishmaref, moving buildings across the frozen lagoon to a

A portion of the giant storm surge barrier system of the Netherlands. More than any other society, the Dutch are preparing for a future with a higher sea level. Inhabiting a small country with no high ground to which to retreat, they inevitably favor an engineering approach. (Copyright © Rijkswaterstaat—Adviesdienst Geo-Informatie en ICT [AGI]. Photo courtesy of Deltaworks Online Foundation)

mainland location (at a cost of several hundred thousand dollars per person) would resolve the immediate sea-level-rise crisis. Abandoning everything—homes, villages, islands, and the nation itself and moving to New Zealand—will solve the Tuvalu sea-level-rise problem. As sea level continues to rise, the Venice problem will likely be solved by abandonment a few decades from now even if the gates are in place. By the time the flooding in Venice has reached an entirely intolerable stage, it is likely that many other cities in Italy and elsewhere in the world will be clamoring for help to combat rising ocean levels. Further, huge outlays of funds to move or jack up the historic buildings will most likely be unthinkable or at least not the highest political priority.

In the United States, too, problems related to sea level rise will multiply over the next few decades. Will the country continue to hold its Gulf Coast and Atlantic barrier island shores in place with costly engineering? Will the public stand for the expenditure of funds to save people imprudent enough to build next to a shoreline known to be eroding? At some point, will the protection of Boston, Manhattan, Philadelphia, Charleston, Miami, Galveston, and a dozen other major American coastal cities become a higher priority than the protection of tourist beach communities on barrier islands?

— Chapter 2 —

Why the Sea Is Rising

There is, one knows not what sweet mystery about this sea,
whose gently awful stirrings seem to speak
of some hidden soul beneath.
HERMAN MELVILLE, *Moby Dick*

THE NORTHERN HEMISPHERE winter of 2007–2008 was the coldest since 2001, according to National Oceanic and Atmospheric Administration's National Climate Data Center. This led many of the voices hostile to the idea of global warming to call once again for a blanket repudiation of warming predictions. We are always fascinated by efforts to argue for or against global warming based on temperature data. Very small changes in temperature can make a big difference, and global trends are very difficult to sort out. From our point of view, one doesn't need any temperature data to see that the planet is warming.

We would hate to be a part of any team charged with measuring temperature trends over the last few years or decades. Where do you begin? Daytime highs? Nighttime lows? Summer temperatures? Winter temperatures? What do you do about the significant portions of our planet where there are no data or data of poor quality? Trying to determine if the planet is warming in this fashion seems fraught with peril if not impossible.

Unusual evidence of sea level rise on a bridge abutment in Miami, Florida. The changing position of barnacles and oysters on this abutment over a 32-year time span indicates an approximate 6-inch (15 cm) sea level rise. Since the last photo was taken, the species on the abutment have changed and thus the sea-level-rise record has not continued. (Photo courtesy of Hal Wanless)

You don't need to measure thousands of temperatures to find evidence on whether or not the planet is warming. Earth does the averaging for us. There are many physical and biological characteristics of our planet that have been responding to long-term changes in average temperatures. For example, studies from both hemispheres indicate that the world's alpine glaciers are retreating. Glacier National Park in Montana is down to 26 named glaciers from 150 in 1850, and if this trend continues, the park is expected to be ice-free by 2030. Will the name of the park have to be changed?

Mount Kilimanjaro's once massive Furtwängler Glacier, made famous by the Hemingway short story, will likely disappear by 2025. Glaciers in the Himalayas are shrinking so rapidly that the flow of

Mt. Kilimanjaro, located close to the equator, has become one emblem of the concern for the impacts of global warming. If current trends continue, Africa's only glaciated peak will be ice-free within a decade or two. (Photo courtesy of NASA)

the major rivers (Ganges, Yellow, Yangtze) they feed may be seriously affected. Permafrost regions are thawing in high northern latitudes, causing buildings to sink, roads to crumble, and a wide variety of other troubles for human infrastructure. And, much has been made of the recent thinning and disappearance of Arctic sea ice. Open water has been reported at the North Pole for the first time since scientists have been able to monitor those conditions. Will Santa have to relocate his workshop?

Then there is the reason for this book: global sea level is rising. This fact is clear from examination of rising tide gauge data the world over and from the increase in coastal erosion along many of the planet's shorelines. So, you don't have to calculate average global temperatures to look for evidence that the planet is warming; its effects are all around us.

In this chapter, we examine the array of significant factors that control the amount of water in the oceans and the level of the sea. Even though it is clear that global warming is the primary driver of modern increases in ocean volume, we will not rehash the debate over whether or not Earth is warming. Clearly, it is.

Ocean Origins

Earth is truly a water planet. Water covers 71 percent of the surface, mostly in oceans and other large water bodies, with 1.6 percent of water belowground in aquifers and 0.001 percent in the air as vapor. Today, the oceans cover approximately 140 million square miles (360 million km²) and average 3 miles (5 km) in depth. In the past, water covered even more of Earth than it does today.

The oceans gave birth to life and nursed that life along for more than three billion years before releasing it to the continents. The early planet lacked an atmosphere hospitable to life, and thus the ocean was life's nursery. Among the earliest life-forms were single-celled plants that breathed oxygen into the atmosphere through photosynthesis (the process through which plants make food). By providing an oxygen-rich atmosphere complete with ozone to block out harmful radiation, the oceans, then, also prepared the planet for life on land. Through complex circulation patterns, the oceans also play a critical role in moving heat from the equator to the poles, allowing the planet to cool itself.

Most scientists believe that the earth formed very hot and dry. Any water existing on this early, scorching world would have boiled off into space. As the planet cooled, water could exist as a gas in the atmosphere and, eventually, as liquid water on the surface. But, if there were no water to start with, where did the 322 million cubic miles (1.37 billion km³) of ocean come from?

There is still some debate. Some of our water was breathed out in vapor form by the planet itself through volcanic eruptions—a process called outgassing. But, this couldn't have been the primary source for the tremendous amount of water found on the planet. Current scientific consensus holds that the rest of the water came from space via ice-laden comets and certain asteroids that crashed into Earth. This water would have arrived very early in Earth's history, when our solar system was loaded with objects left over from the formation of the planets. The fact that most of our water is of cosmic origin is a wonderful thing to ponder.

By somewhere around four billion years ago, the earth's oceans had formed, and the earliest life-forms, simple single-celled organisms, arose shortly thereafter. This early planet looked quite different from our planet of today. Any continental landmasses would have been very small. Earth was dominated by the newly formed global sea. Since that time, the area of the planet covered by continents has increased and the total area of the oceans has decreased, though the amount of water on the planet has remained essentially stable.

Dynamics of Continents and Oceans

Imagine, for a second, removing all of Earth's surface water and emptying the oceans. You would find a planet with very large, deep basins separating the much higher continental areas. These basins were not carved out of the earth, like a river might form a canyon; rather, they are a fundamental result of plate tectonics, a process that plays an important role in sea level change.

The ocean basins are relatively low in elevation because the rocks that form the crust beneath the oceans are different from the rocks that form the crust on the continents. An understanding of plate tectonics tells us that the earth is made up of layers, with the upper layer, the lithosphere, sitting atop a layer called the aesthenosphere. The lithosphere (which includes the crust and upper mantle) is rigid and relatively cool. The aesthenosphere is relatively warm and ductile. This difference allows the lithosphere to "float" on top of the aesthenosphere.

In simple terms, Earth's lithosphere is made up of two basic kinds of rocks. The thinner lithosphere underlying the oceans is comprised predominantly of rocks called basalt and gabbro, rocks that are relatively dense compared with rocks such as granite and rhyolite that make up the thicker continental lithosphere. This density allows the oceanic lithosphere to sink down into the ductile aesthenosphere.

Ocean basins exist wherever the dense, thin lithosphere is riding low in the aesthenosphere. Continents exist wherever the relatively light, thick lithosphere is riding high on the aesthenosphere. Most of the world's water collects in the lithospheric low spots because, well, water runs downhill. So the contours of the earth's oceans were not carved; rather, they exist simply because of differences in lithospheric composition, which in turn is explained by the action of plate tectonics.

The plates of "plate tectonics" are sections of lithosphere approximately 62 miles (100 km) thick, giant moving puzzle pieces that make up the earth's surface. These plates are in constant motion, sliding on top of the partially liquid aesthenosphere. As they move, they bump into each other, slide past each other, and sometimes pull away from each other. Plate boundaries are where the action is, geologically speaking. It is here that we find most of the world's volcanoes and earthquakes. Through a variety of complex interactions, plate movement can also form new lithosphere and consume older lithosphere. Plate movement can also push up mountain ranges where the tectonic plates collide (e.g., the Himalayas), open up ocean basins where the plates spread apart, and affect long-term sea level rise.

Interestingly, once continental lithosphere is formed, it is very difficult to destroy, whereas the very dense oceanic lithosphere sinks back down into Earth's aesthenosphere with relative ease (a process called subduction). When continents are formed, then, they are essentially here to stay, though their shorelines may fluctuate with rise and fall in sea level. Our planet has thus been gaining continental landmass at the expense of oceanic lithosphere since its formation, but the process of continental gain is very slow, and getting slower as Earth's interior cools.

Sea Level and Its Rise

Before describing the factors that change sea level, it is helpful to say a little about what *sea level* means. Most simply, sea level is the average height of the sea with respect to some reference surface. Most of us think of sea level as being the 0.0-foot elevation contour. In reality, most elevations on land are referenced not to mean sea level but to a standardized reference point used by surveyors called a *geodetic datum*. As a result, the surveyed "elevation" of sea level is almost never 0.0 feet.

Determining the average height of the sea involves sorting through a variety of regular water level fluctuations. While this book focuses on long-term changes in sea level, such as those resulting from global warming, there are many short-term fluctuations in sea level that complicate a simple measurement of average sea level (or mean tide level).

In order to measure subtle changes in long-term sea level, we have to compensate for water level fluctuations driven by waves, tides, currents, storm surge, atmospheric pressure differences, and ocean surface topography resulting from large-scale ocean circulation. Fortunately, this is typically a fairly straightforward process as most of these fluctuations are regular and can be easily separated from the sea-level-rise signal. To make measures of sea level even more interesting, the seas, it turns out, are not actually level. The gravitational pull on ocean water varies across the surface of the earth due to variations in the diameter and composition of the planet. If the oceans were allowed to sit completely still, the water would come to rest at an elevation in equilibrium with that variable gravitational attraction. This theoretical still water surface is called the *geoid*. It is an equipotential surface—a surface where the gravitational attraction is equal at all points. Even though the geoid (and thus the ocean surface) has approximately 656 feet (200 m) of relief, an object traveling along the geoid will weigh exactly the same at any point on the geoid, whereas if you carried that object up a mountain on land, it would weigh less as the gravitational pull decreased.

A sailor traveling from New York across the Atlantic will actually sail uphill tens of meters (with respect to a fixed reference point in space) on her way to Great Britain. But she won't notice the change, just as we don't feel like we're upside down when we visit the Southern Hemisphere (or vice versa for our friends from south of the equator). The gravitational pull is always perpendicular and always equal.

Because the geoid is related to the interior density of the earth, it tends to be relatively stable compared to changes in sea level due to global warming. But there have been instances when sea level along one stretch of shoreline has changed as a result of fluctuations in the geoid, driven by tectonic changes within the earth. Geoidal fluctuation has been suggested as the driving mechanism in an unusual rise in sea level along the coast of Brazil a few thousand years ago. In general, though, predictions of global sea level change over the next century do not need to take the geoid into account because geoidal fluctuations operate on very different timescales.

In broad terms there are two other types of sea level change—local (isostatic) and global (eustatic)—that need to be taken into account. Isostatic changes in sea level result from factors that are causing the local lithosphere to move up and down. These factors include mountain building resulting from tectonic plate collisions, glaciation, sediment compaction in large deltas, or anything that adds to or subtracts from the weight of the crust. For example, the rate of sea level rise in Louisiana is very different from the rate of sea level rise just down the coast in Alabama. The difference is due to local subsidence of the Mississippi River Delta in southern Louisiana. There are many places at high latitudes where sea level is in fact currently falling. This is because the land is rising as glacial melting removes weight from the lithosphere. Isostatic, or local, relative sea level rise, then, can vary greatly from place to place.

Eustatic sea level rise is simply a measure of the increase in the volume of water in the oceans, expressed as a change in the height of the oceans. Currently, eustatic sea level is rising at a rate of approximately

0.08 inch per year (2 mm/yr) if averaged over the last hundred years and around 0.12 inch per year (3 mm/yr) over the last fifteen years. The rate of eustatic sea level rise is accelerating in response to global warming.

There are two ways in which global, or eustatic, sea level change can occur: (1) change in the accommodation space in the ocean basins brought about by a change in the rate of tectonic plate movement and (2) change in the volume of ocean water. Modern sea level rise is being driven by the latter, increases in volume, but in times past the former has generated much more dramatic excursions of the ocean onto land.

Oceans on Land: Sea Level Change in the Geologic Past

Walking around the Flint Hills of Kansas, one can find a wonderful diversity of fossils. These fossils are all marine organisms from the Permian period, the geological interval that stretched from about 300 million years ago to roughly 250 million years ago. Although the fossils represent organisms found in the ocean during that period, it isn't exactly correct to say that Kansas was an ocean during the Permian. It is more correct to say that the area that is now Kansas was once covered by an epeiric sea.

Epeiric seas form when sea level rises dramatically, on the order of hundreds of meters, and the oceans spill out of the ocean basins and up onto the continents. Epeiric seas form as a result of acceleration in the rate of tectonic plate movements and a concomitant change in the shape of the ocean basins. When the rate of plate tectonics is rapid, the oceanic lithosphere rides higher on the aesthenosphere, leaving less room for the same volume of water in the ocean. The ocean basins simply can no longer hold all the water, and the seas flood the lower portions of the continents. Most of the world's coal deposits can be found at the edges of former epeiric seas where large continental wetlands once existed. The last, great epeiric sea receded from the continents more than sixty-five million years ago as the rate of plate tectonic movement slowed. Changes in the rate of plate movement occur on timescales much too long to have any impact on the decadal-scale changes we are experiencing today.

Changing Volumes of Ocean Water

The sea level rise of current concern is a result of a change in the volume of water in the ocean basins rather than a change in the size of the basins themselves. The volume of water in the oceans can be increased in two ways: by the addition of more water or by heating the water. Both of these factors are fairly straightforward as concepts.

Water expands as it warms: heat 50 gallons of water to 100 degrees Fahrenheit and you will have roughly 51 gallons. As the planet's atmosphere warms, the ocean absorbs a tremendous amount of that heat. Thermal expansion of ocean water is expected to continue to play an important role in rising sea level throughout the next century. The most recent rise estimate by the United Nations Intergovernmental Panel on Climate Change (IPCC) is based primarily on thermal expansion.

To understand how to add water to the oceans, it helps to think back to grade school and the first simple description given in science class of Earth's water cycle: water evaporates from the oceans, falls back to Earth as rain, and flows back into the oceans through rivers (more or less). But your grade school teacher probably forgot to tell you about a wrinkle in this scenario—glaciers. If climate is getting colder, more of that precipitation will fall as snow and get stored in glaciers. If climate is getting warmer, the glaciers will begin to melt and return water to the sea. So, fluctuations in global temperature will change the balance between water stored on land and water stored in the oceans.

Currently, nearly all of the world's glacial ice is retreating and thinning, adding water to the oceans. The feared collapse of the Antarctic ice sheets and melting of the Greenland ice sheet could create a catastrophic rise in sea level of as much as 204 feet (61 m). The health of the world's glaciers will be closely watched by scientists monitoring the future of rising sea levels.

Scientists have developed an excellent understanding of how the coming and going of glaciers has changed sea level over the last million years or so. During the peak of the last ice age, around twenty-one thousand years ago, global sea level was approximately 400 feet (120 m) lower than it is today. The water "missing" from the oceans then was stored in the large continental ice sheets that covered much of northern Europe and North America. This may seem like a long time ago, but our fellow *Homo sapiens* were around to see it.

Multiple lines of evidence confirm this sea-level-low stand. There are submerged shorelines on the continental shelves off many coasts. There are old river channels, now submerged, reaching to the edge of the continental shelf. Fishermen have dredged up artifacts of coastal settlements, and Dutch fishermen regularly dredge up mammoth bones from the North Sea near Scandinavia. Old coral lie abandoned by rising sea level offshore of many tropical islands.

In fact, ice ages have come and gone approximately every hundred thousand years for the last six hundred thousand plus years. The warm periods between ice ages, like the one we are experiencing now, are called interglacials. During some of these warm periods, sea level was even higher than it is today. In Bermuda and the Bahamas, old shorelines now visible on land indicate that sea level rose as much as 13 feet (4 m) higher than it is today at least three times in the last half million years. Along the U.S. East Coast, relict barrier islands were stranded as part of the coastal plain when sea level dropped during the last ice age. In many cases, these abandoned islands have now been intercepted by the modern rise in sea level, and new barrier islands have formed adjacent to the old shoreline, such as in the state of Georgia.

Obviously, all of these ups and downs in ocean volume happened without the human interference that is being blamed for an acceleration of sea level rise today. So what caused these ice ages and warming periods that so dramatically affected climate and ocean volume? Although there is still some scientific debate, it is generally agreed that it was some combination of changes in Earth's orbit around the sun, changes in the composition of the atmosphere, and changes in ocean circulation—all of which may be causally linked in ways that we are yet to understand.

Continuous ice core records from Greenland and Antarctica provide a peek into the composition of the atmosphere during these past climate shifts. These cores contain trapped air bubbles preserving atmospheric gases stretching back 650,000 years. These data indicate that greenhouse gases fluctuated up and down along with global sea level as the ice ages came and went. This might seem to remove humans as a possible culprit for today's global warming. Clearly, the concentration of carbon dioxide in the atmosphere fluctuates naturally.

But these records serve primarily to put our current climate crisis in perspective. Current atmospheric concentrations of greenhouse gases like carbon dioxide and methane far exceed levels seen in any of the climate records dating back 650,000 years. The concentrations of these gases took off after the Industrial Revolution. Based on these data, the IPCC declared that it is *very likely* that humans are driving an acceleration of global warming. *Very likely*, to the IPCC, means greater than 90 percent probability. It is worth noting that the IPCC works by scientific consensus and thus likely is very cautious in its estimates.

It is clear that climate has warmed naturally since the end of the last ice age, and sea level has risen substantially. But now human activities may be pushing the warming and the associated rise in sea level at a rate and to a level not seen in the recent past. And it is important to remember that during the past fluctuations of sea level, there were no places like New Orleans, Bangladesh, the Nile Delta, Venice, or New York City to worry about.

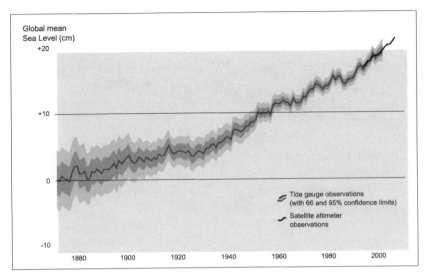

This graph shows a global average of reliable tide gauge data from 1870 to 2000 with the new satellite altimetry data superimposed on the end. There has been a clear acceleration in the rate of sea level rise since 1990. Current tide gauge data and sea-level-rise numbers from satellites correspond closely. (Photo courtesy of NOAA)

Measuring Modern Sea Level Rise

The fact that sea level is rising is not debated to the same extent that global warming deniers debate the warming of our planet. The evidence for sea level rise is clear. The United States spends hundreds of millions of dollars each year managing beaches that are eroding with at least some contribution from sea level rise. Venice is flooding. The Netherlands is walled off from the sea.

Even more convincing, sea level rise can be directly measured in a far more straightforward way than something like global temperature. The longest record of direct measurement of sea level comes from tide gauges. A tide gauge is a device built to measure water level variations due to tides and weather and to eliminate effects due to waves. A tide gauge can be as simple as a long ruler nailed to a post on a dock. More sophisticated instruments are usually placed in a stilling

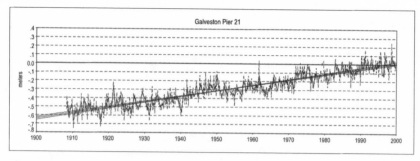

A 100-year record of tide gauge levels from Galveston, TX. This area is experiencing subsidence, so the relative rate of sea level rise is higher than the global average for eustatic sea level rise. (Photo courtesy of NOAA)

well, or pipe, that protects a float connected to a recording device from waves. As tides rise and fall, the float's motion is recorded.

Tide gauges were not built with the intention of measuring changes in sea level, but they have proven very useful for doing so. A glance at a simple tide gauge record like the one from Galveston, Texas, reveals a complex history of water level change with a clear upward trend. The Galveston record is typical of most of the world's datasets, with the primary differences from place to place being the slope of the line (the rate of sea level rise or fall) and the duration of the dataset. When looking at a tide gauge record, remember that the data are representative only of local relative sea level rise (as discussed above). So areas that are experiencing tectonic or isostatic change will bias any attempt to determine the global, eustatic sea level signal. The Galveston record shows an average rise rate of 0.3 inch per year (7.6 mm/yr), for example, but this coast is subsiding, so the gauge provides an overestimate of global sea level change.

The way to get around this is to examine many tide gauge records from all over the world, especially ones that cover a significant period of time and have been drawn from relatively stable coasts, such as the waters along the coast of Florida. The tide gauge record from Fernandina Beach, Florida, for example, indicates a rate of rise of approximately 0.08 inch per year (2 mm/yr) since about 1900, when recordings were first taken. These rates from Florida are remarkably close to the global average rate of around 0.07 inch per year (1.8 mm/yr) that was determined by averaging tide gauge records of good quality from stations all over the globe.

Sea level can also be measured from space, and the data derived from doing so closely parallel that from tide gauge measures. The TOPEX/Poseidon satellite, launched in 1992, measured sea level and was also used to map the ocean floor. As TOPEX/Poseidon orbited Earth, an altimeter bounced radar signals off the ocean's surface. The altimeter recorded the time it took for the radar signal to return to the satellite, and that gave a precise measurement of the distance between the satellite and the sea surface. Measurement from orbit is the only way to assess sea level independently of land-level changes. TOPEX/Poseidon served as the primary means for monitoring the oceans until it was supplanted by the more accurate sea level monitoring satellite Jason-1, launched in 2001.

The global average rate of sea level rise measured via satellite altimetry for the period 1993–2003 was 0.12 inch per year (3.0 mm/yr) according to the IPCC. Interestingly, the eustatic sea-level-rise signal is not uniform in the global ocean. Variability in ocean temperatures, ocean currents, and events like El Niño can locally mask the global signal. Sea level in the Atlantic and the eastern Indian oceans was rising during 1993–2003, while sea level in the eastern Pacific was falling in that period, for example.

A 2007 IPCC report estimates that for the period 1961–2003, approximately 60 percent of the sea level rise was due to an addition of freshwater to the oceans from melting glaciers, while 40 percent was due to thermal expansion. For the period 1993–2003, the ratio reversed, with thermal expansion accounting for 60 percent of the rise. These data may indicate why coastal erosion seems to have reached crisis levels along many of the world's shorelines over the last two decades. Even communities that have been around for a while, like Happisburgh on England's east coast, are seeing dramatic losses of land and property.

Sea level has clearly been rising at an accelerating rate through the twentieth century and into the twenty-first. The real issue, which we take up in the next chapter, is whether scientists can accurately predict the future trends of sea level rise.

— Chapter 3 —

Predicting
the Unpredictable

Prediction is very difficult, especially if it is about the future.
NEILS BOHR, Nobel Prize–winning physicist

*I*MAGINE IF you knew the future of the stock market, the course of a health problem, the path of a war, the day on which a major hurricane will strike a particular shoreline reach, or the future course of a love affair. What a huge difference this could make for your future prosperity and well-being.

Our society invests a lot of effort and resources in attempting to predict these events (except for love affairs, of course). An army of prognosticating experts in well-tailored, dark suits chant the future of the stock market and explain away unexpected drops or surges in stock prices on all the TV news channels, while politicians, retired generals, and other armchair analysts foretell the outcomes of wars in optimistic or pessimistic terms, depending on their political slant.

Recent events on the economic front have certainly not helped the credibility of stock market prognosticators. In fact, the deteriorating value of the stock market in early 2009 inspired Warren Buffet to warn, "Beware of geeks bearing formulas."

Now imagine that we could accurately forecast the precise future extent and timing of global sea level rise. If such a prophecy were taken as a certainty, it would precipitate a lot of immediate action. Coastal cities would start planning new storm drainage systems, begin discouraging redevelopment in low areas, and perhaps even begin to move, abandon, or protect various portions of the city.

Next, assume it could be possible to predict how the rising level of the sea will remake a shoreline. The value of oceanfront property located along sandy shorelines would drop precipitously, and strong new rules to encourage construction of movable buildings and discourage high-rise buildings next to beaches would likely appear.

Predicting the outcome of natural processes, particularly those that are hazardous to humans, has long been a major goal in almost every field of science. Billions of dollars have been spent in attempts to learn when and where the next earthquake, landslide, tsunami, river flood, ice storm, volcanic eruption, hurricane, tornado, drought, or asteroid strike might occur. Since the 1990s, global climate change and its related effects have been added to the top of this list of possible natural hazards. In the United States alone, funds were being spent in 2008 at a rate of $2 billion per year to predict the future of climate change and its impacts, such as sea level rise, despite the George W. Bush administration's frequent denigration of the issue's importance.

The importance of knowing something of shoreline change rates can be illustrated by the story of an upstate New York couple who were interested in purchasing a house on Topsail Island in North Carolina. In 2006, they visited the island, chose, and bought their beachfront retirement dream house, and went back home to settle affairs and get their furniture. Upon return, they found their new house was standing out on the beach and condemned for habitation. An irresponsible real estate agent had sold the property to the couple without mentioning that the house had been condemned once before. An artificial beach had been pumped in after the first condemnation, putting the house temporarily landward of the high tide line—just long enough to be sold to the unwary couple. The North Carolina legislature, as a result of this incident, briefly considered, but didn't pass, a real estate disclosure law requiring that sellers fully disclose the nature and magnitude of natural hazards that could affect a property.

On a larger scale, stakes are high for thousands of high-rise buildings along the low-lying retreating shores of the U.S. Gulf and Atlantic coasts. Which ones will be threatened this year? Next year? Next decade? And since government is frequently financially involved with rescuing such buildings, how much will their protection cost taxpayers?

Predicting the future of nature is not only fraught with difficulty but has long been a politically hazardous venture. Years ago, the government of India contemplated halting weather forecasts because the ruling party was blamed for inaccurate forecasts. More recently, the 2005 failure of New Orleans levees provided a spectacular example of an engineering disaster with heavy political fallout. This was both a predictive failure and a political failure. Scientists had predicted for years that New Orleans was likely to be flooded when the "right" hurricane came by. In April 2009, Ivor van Heerden was removed as director of the Louisiana State University Hurricane Center. His crime was pointing out these failures to the public and the media. To be useful, predictions must of course be heeded; in this case, few local and national authorities or politicians took them seriously. Adding to the problem were poorly engineered levees and pump systems, which meant that damage was unpredictably extensive.

How do scientists predict the outcomes of natural processes? As a first cut, it is done on the basis of past experience. All would agree that Los Angeles is more likely to suffer an earthquake than is Chicago, based on past experience in these localities and on predisposing geologic conditions. We also know from experience that Louisiana is more likely to suffer damages from a hurricane storm surge than is coastal Maine. As a second cut, prediction is done on the basis of field studies of the particular process in question. For example, satellite and on-site field observations have revealed an ominous increase in degradation of the world's ice sheets at a rate that is likely to lead to a large sea level rise.

In recent years, mathematical modeling has become the primary tool used to fine-tune predictions. The widespread availability of powerful desktop computers, new government requirements for environmental impact statements and cost-benefit calculations, and increased confidence among scientists and engineers that predictive mathematical models would be quite accurate—all arrived on the scene simultaneously in the final quarter of the twentieth century.

A mathematical model is simply a description of a natural process with mathematical equations. Scientists in the 1960s and 1970s anticipated that quantitative mathematical models would at last provide the long-hoped-for basis to predict accurately the future of natural processes. Almost overnight, mathematical modeling became the embodiment of sophistication and state-of-the-art in science. Ever since, mathematical modeling has held sway in the nation's effort to predict the future level of the sea.

The use of mathematical modeling for predicting the outcome of important societal issues has been problematic. When Hurricane Floyd passed by Charleston, South Carolina, in 1999, weather prognosticators predicted it would go ashore over the city. It didn't, but many thousands of South Carolinians who were ordered to evacuate found themselves trapped all night long, immobile in cars containing family photos, children, cats, dogs, and goldfish, on a jammed Interstate 26 leading out of the city. Considerable public anger was expressed, not over the inaccurate hurricane alarm (because everyone understands the inaccuracy of storm path prediction), but because local planners incorrectly assumed that Interstate 26 had sufficient capacity to evacuate the city. In fact, the road may well have been adequate, but unexpectedly, the state's governor overruled emergency management officials who had planned on making all lanes of the interstate outbound from the city to accommodate the exodus.

Governor Jim Hughes's inexplicable decision in a time of crisis

Attempting to escape from Charleston, South Carolina, before the arrival of Hurricane Floyd in 1999. Governor Jim Hughes decided to keep traffic lanes open in both directions. It was a colossal mistake. Emergency management officials who had modeled efficient storm evacuation had not imagined the governor would make such a decision--one instance of how difficult it is to model human behavior. (Photo courtesy of the U.S. Army Corps of Engineers)

is an example of the problem of predicting anything that involves human behavior (as in the botched evacuation of New Orleans during Hurricane Katrina and the incompetence of FEMA, the Federal Emergency Management Agency). Emergency management officials in Charleston had carefully studied and modeled hurricane evacuation long before the storm. But the models didn't include the possibility that the governor would be convinced by some of his aides that the incoming lanes should be kept open for "emergency vehicles."

More recently, Paul Volcker, the eighty-one-year-old special economic adviser to President Barack Obama, also explained the 2008 Wall Street market breakdown as a failure to recognize that human behavior is not predictable. He noted that "no mathematical model can accurately predict human hysteria in a financial panic."

Just like predicting the success of Wall Street futures or escape from Charleston, prediction of the future of sea level rise depends on accurately predicting human behavior, albeit on a much larger scale. The extent of sea level rise in the next century is in part dependent on the future abundance of greenhouse gases, which is dependent on human behavior. But who knows how much carbon dioxide our children's and our grandchildren's generations will produce or, for that matter, how much our own generation will produce over the coming years? When might we develop the political will to significantly reduce carbon dioxide and other greenhouse gas emissions? What energy technology might we humans devise that could largely supplant fossil fuel? How will human-induced changes feed into the natural background changes in climate? How can we accurately predict the course of global warming without this knowledge? These questions, unanswerable in fine detail, are addressed through assumptions made in the mathematical models that we discuss below. And therein lies a major problem with mathematical models: the uncertainty of basic assumptions.

Prediction comes on many scales. Weather forecasts are often accurately projected two to five days in advance. Although the success rate for short-term weather prediction is relatively high, each of us has been caught without an umbrella on an unexpectedly rainy day. The possibility of increasing the time span of equally accurate weather prediction is slim: beyond two to five days out, atmospheric changes that will affect weather are often unknown and unpredictable.

Climate is the average state of the weather over a period of years. Predictions of future climate are of course more general than predictions of the next day's weather. You would not want to rely on climate predictions to decide whether to bring an umbrella to the office. You might want, however, to purchase a well-built house situated a good distance back from the shoreline and at high elevation if climate predictions indicated more frequent or more powerful hurricanes were on their way in coming decades.

A predictive mathematical model provides a forecast of the end result of a natural process described by an equation(s). Broadly speaking, there are two kinds of mathematical models, quantitative and qualitative, and the distinction between the two is very important to understand. Quantitative models produce a number. Qualitative models produce an understanding. *Quantitative* models are assumed to provide accurate projections into the future, usually expressed as a single number or a small range of numbers. In quantitative modeling, all the important parameters that affect a process must be understood and expressed in numbers. No important parameter can be omitted or ignored. Such models usually answer the questions *where, when,* and *how much.* Where will a storm strike the shoreline? When will the melting ice sheets reach an irreversible tipping point? How much will sea level rise?

Qualitative mathematical models are also based on mathematical equations and are sometimes called process models or scenario models. These do not provide specific numerical predictions but are used to understand processes or to ascertain trends, order-of-magnitude answers, or answers with large plus or minus error bars. Only an estimate of the most important parameters is needed for qualitative modeling, and although the answers may be very useful, they are not assumed to provide accurate numbers, only generalizations or trends. *How, why,* and *what if* questions are answered. How will reduction of CO_2 emissions impact the sea level rise? Why is the Arctic Ocean sea ice melting? What if Antarctic ice shelves continue to collapse?

Quantitative models might predict (as a hypothetical example) a sea level rise during the next century of 3.3 to 3.9 feet (1 to 1.2 m) and a local shoreline retreat of more than 900 feet (270 m) in the same period. Qualitative models in the same arena might predict that the sea level will continue to rise and might accelerate and that the retreat of the shoreline will continue and probably accelerate as well.

Qualitative models have provided quite useful scenarios for thinking about our future climate and the level of the sea. But we can't plan for the future of sea level rise if we simply predict that sea level will rise. Numbers are needed; estimates, even if rough, are critical. How else can planners design future drainage systems for cities? How else to plan the future of barrier islands, the future of coastal cities, the size of storm gates protecting Venice, London, and Rotterdam?

In reality, scientists and planners use a combination of qualitative and quantitative models, field observations, and studies of the past behavior of ice sheets and come up as best they can with sea-level-rise estimates with large pluses and minuses. Field observations are most important of all. For example, recent field and satellite studies of the Greenland and Antarctic ice sheets have significantly raised the expected rate of sea level rise in the twenty-first century. Mathematical models didn't help in this regard, as we'll see in the next chapter. And after all is said and done and a number has been agreed upon for engineering design purposes, it is critical to remember the fragility of the numbers and to be prepared to be flexible, to change horses in midstream.

Studies of ice sheets from the past and how they evolved during the great ice ages have also helped immensely in establishing what can be expected from the current phase of ice degradation. This is especially true of sea level rise, because combined evidence of the ages of old shorelines left behind when the glaciers melted and their elevations (or depths) from all over the world has allowed scientists to construct sea level curves. These are graphs plotting the level of the sea versus age or time, and they give some indication of the rate at which sea levels changed in the past and how they might change in the future. These rates provide boundaries of reality for the future of sea level rise.

Predicting Sea Level Rise

The United Nations Intergovernmental Panel on Climate Change (IPCC) produced its fourth assessment (the AR4 report) on global change in 2007. It is one of the most important documents of our time, and for their work in developing the report the scientists who produced it received the 2007 Nobel Peace Prize. The report was published in three volumes: (1) The Physical Science Basis; (2) Impacts, Adaptation and Vulnerability; and (3) Mitigation of Climate Change. Much of the predictive modeling of sea level rise is reported in volume 1.

The reports are thick and dense in exposition. A real effort is required to understand and interpret them. Even the brief summary document the IPCC produced for policy makers and nonscientists is complex. Probably what makes the reports difficult is that the writing is done or approved by a wide array of scientists with their characteristic reserve, reticence, and insistence on highly qualified statements. Still, all in all, these are amazing documents produced by a committee of twenty-five hundred!

The following conclusions concerning sea level rise are paraphrased from the Summary of AR4 for Policymakers.

I.   The level of the sea will rise between 7 and 23 inches (18 and 58 cm) by the end of the twenty-first century due mostly to ocean warming and consequent expansion of its waters plus melting of mountain glaciers. Thermal expansion of the oceans will continue for many centuries, even after greenhouse gases are stabilized (i.e., reach a peak and begin to go down). Sea level will rise 0.7 to 2.0 feet (0.2 to 0.6 m) due to thermal expansion per degree Celsius of global average warming.

II. An additional amount of sea level rise will result from "melting ice sheets, glaciers and ice caps" of the Greenland and the West Antarctic ice sheets. Recent field observations have indicated that the contribution to sea level rise from both of these ice sheets may be much larger than previously assumed. Most likely, such melting will itself produce a rise in sea level larger than that due to ocean expansion. "Rapid sea level rise on a century scale cannot be excluded."

Fuzzy and complicated wording of IPCC documents has led to large misunderstandings and created much fodder for global warming opponents. The media now widely accept that the predicted sea level rise will be between 7 and 23 inches (18 to 58 cm) by century's end, but as just noted, this range doesn't include critical, perhaps catastrophic, increases due to ice sheet melting.

Another widespread but flawed interpretation is that the 2007 report predicts a lower maximum sea level rise than the 2001 IPCC report did—down from 35 inches to 23 inches (89 cm to 58 cm). As Stefan Rahmstorf of the Potsdam Institute for Climate Impact Research points out, however, the earlier, higher estimate includes the expected contribution from the melting of Greenland ice but the 23-inch (58 cm) maximum of the 2007 report does not, thus making the comparison one of apples and oranges. How would anyone except deeply involved scientists who studied both of these dense reports know that, though?

These confusions are a result of the IPCC's decision for its AR4 report to provide actual predictions for only some of the causes of sea level rise (those it could predict with mathematical models). It produced predictions based on sea level rise from thermal expansion and mountain glacier melting while refusing to assign numbers—because committee members didn't have a mathematical model they agreed on—to what is likely to be the most important source of sea level rise in the twenty-first century: ice sheet melting.

Gary Mitchum, University of South Florida expert on sea level rise, explains that within the scientific community the assumed importance of Greenland and Antarctica has evolved with time. In the 1970s and 1980s, both masses of ice were considered by global change experts to be potential sources of rapid sea level rise. This changed in the 1990s when the general consensus was that the impact of global warming on ice sheets would be quite gradual and that, of the two ice masses, Greenland's would be most important. But in the last few years both Greenland and the West Antarctic ice sheet have been resurrected as potential sources of water for a disastrous increase in the rate of sea level rise.

Dyed-in-the-wool modelers unfortunately have evolved to the point where, without a model, they deem prediction impossible or too inaccurate to be useful. Yet we need some sort of thoughtful prediction of the role the ice sheets will play in future sea level rise, and if experts can't do it, who can? At least an educated guess is needed. In its defense, it should be remembered that the IPCC is a very large committee whose members worked in a politically charged atmosphere, so such an estimate may have been far beyond what their consensus process could achieve.

In the absence of sufficient guidance from the IPCC, a number of national, state, and local organizations in the United States and elsewhere have begun to make their own predictions. One of these is the Science and Technology committee of Florida's Miami–Dade County Climate Change Task Force. Committee members argue in their January 2008 report that the *minimal* figures for sea level rise over the course of the twenty-first century are as follows.

Thermal expansion of ocean water . . 11.8 inches (30 cm)
Mountain glaciers . . . . 3.9 to 9.8 inches (10 to 25 cm)
Greenland ice sheet melting . . . . 18.9 to 33 inches (48 to 84 cm)
Antarctica ice sheet melting . . . . 3.9 to 5.9 inches (10 to 15 cm)
Minimal total . . . . 3.3 to 5 feet (1 to 1.5 m)

Rhode Island's Coastal Resource Management Council independently came up with basically the same numbers as the Miami committee: a minimal rise of 3 to 5 feet (90 to 151 cm) by the year 2100. The U.K. Climate Change Impacts Programme suggests that a minimum sea level rise of a little less than 3 feet (0.9 m) should be expected or at least assumed by 2080 and an 8-foot rise (2.4 m) should be considered a possible extreme. In a 2008 paper in the journal *Science,* University of Colorado scientist Tad Pfeffer and eight colleagues argue that sea level rise could range from slightly less than 3 feet to a maximum of 6.5 feet (less than 0.9 to 2.0 m) by 2100.

In a 2007 *Science* article, Stefan Rahmstorf and six colleagues compared past IPCC model projections with actual sea level rise both from tide gauges and satellite measurements from 1990 to 2006. When they took the difference between predicted and actual sea level rise during that period, they projected that the actual sea level rise between 1990 and 2100 will likely be higher than the IPCC estimates. Their calculations suggested that sea levels in 2100 could be 1.6 feet to 4.6 feet (0.5 to 1.4 m) above 1990 levels.

Rahmstorf and associates' projection is considered a Business as Usual (BAU) approach to climate change and sea level rise because it assumes linear change, simply a straight-line projection of processes of the last century into the next century. James Hansen of NASA's Goddard Institute for Space Studies emphasizes that thermal expansion and mountain glacier melting dominated sea level rise in the twentieth century, but ice sheet disintegration will become the most important component of the twenty-first century and will be non-linear in its course. That is, the change in sea level versus time will likely not plot as a straight line on a graph. The contribution of water from the world's great ice sheets has doubled since the 1990s, which Hansen believes is a strong basis for assuming that a much higher rise in sea level than Rahmstorf's estimate in the next hundred years is a real possibility:

> Let us say that the ice sheet contribution is 0.4 inches (1 cm) for the decade 2005 to 2015, and that it doubles every decade until the West Antarctic ice sheet is largely depleted. That [rate] yields a sea level rise of the order of 16.4 feet (5 m) this century. Of course I cannot prove that my choice of a ten year doubling time for non linear response is accurate, but I am confident that it provides a far better estimate than a linear [BAU] response for the ice sheet component of sea level rise.

Meanwhile the IPCC prediction of sea level rise, ignoring the contribution of the ice sheets, continues to confuse. For example, a November 11, 2008, *New York Times* article reporting on the Maldives plan for survival noted that according to the IPCC, the Maldives are threatened by a sea level rise of perhaps 2 feet by 2100. Clearly, the IPCC report has diminished the understanding of the actual sea-level-rise threat, even at one of the world's leading newspapers.

Predicting Shoreline Retreat

The rise in sea level will affect natural environments and human infrastructure in many ways, some of them catastrophic. In the absence of humans, however, nature at the shoreline would fall back and rebound with relative ease as global changes in sea level transpire. Nature has been responding to changing levels of the sea for billions of years. Without human interference, beaches would always be at the shoreline, albeit in a different location as time rolls on. Barrier islands would migrate in a landward direction, keeping pace with the rising sea (provided it didn't rise too fast). Salt marshes and mangrove forests would move inland as the intertidal zone moves inland and occupy the former floors of forests that fringed the retreating shoreline.

It won't be as easy for humans, however. Buildings can't migrate like barrier islands. Roads, water, surface runoff drainage, sewer lines, and power lines are not in the least flexible. By some predictions, environmental refugees, such as the people who will need to flee the lowlands of Bangladesh, the Niger and Mississippi deltas, or the mid-Pacific atolls with rising waters, are likely to far exceed the number of political refugees in the coming century. In most cases, local property owners will insist on holding barrier islands in place as long as possible to prevent them from migrating. Seawalls will abound and grow in size, and beaches will be degraded, perhaps replenished for a while, but eventually lost. Tourists at the coast will do the Cape May, New Jersey, thing: promenade on top of the seawall with its beautiful view of the sea at high tide and crowd onto the narrow pocket beach exposed only at the lowest of tides.

Along most of the world's coasts, shoreline retreat is already occurring. For example, the rate of retreat of most barrier island ocean shorelines along the eastern coast of the United States already ranges between 2 and 4 feet (0.6 and 1.2 m) per year. Clearly, accurate prediction of the speed of the coming shoreline retreat in the years ahead is a high priority for many people.

If the sea level rose 20 feet (6 m) tomorrow, the slope of the land surface would determine the location of the new shoreline and predicting the location of the new shoreline would be a simple matter. In this imaginary scenario, after an overnight 20-foot (6 m) rise, the new shoreline would correspond exactly to the 20-foot map contour before the water rose. But currently we are not dealing with an overnight 20-foot sea level change. Instead, we are faced with an annual rise more in the range of 0.12 inch (3 mm).

Adding to the complexity of predicting shoreline retreat is the huge variety of shoreline types all over the world. Some are sandy, like the barrier islands of the American East and Gulf coasts, the coasts of southern Spain and Portugal, and long reaches of the Arctic Ocean shoreline. Others, like the shorelines of the Aleutians, Japan, the Caribbean, and the Philippines, are hard volcanic rock. Many coasts, especially the mountainous west coasts of North and South America, are lined with shores made up of many types of rock with varying degrees of resistance to the forces of breaking waves. In the Arctic, beaches are frozen and the rate of erosion depends in part on how much of the permafrost melts each summer. On sandy, temperate zone beaches, compacted mud, peat, and rock layers are often present, strongly affecting rates of shoreline retreat. How can one possibly predict the future impacts of a gradual sea level rise on the rates of shoreline erosion on such a variety of coastal slopes and coastal materials? The answer: accurate prediction is impossible.

Only one mathematical model, the *Bruun Rule,* purports to predict shoreline retreat due to sea level rise, and that is only for coasts that are entirely made up of sand. All the world loves the Bruun Rule. It is probably the world's single most widely used mathematical model to predict the outcome of any natural process. Virtually every coastal nation with even a rudimentary coastal management program has used this rule in one context or another to forecast how rising sea level will affect the location of the ocean shoreline in the future. The general principles behind the Bruun Rule are incorporated into most of the numerical mathematical models coastal engineers use to

An ocean-facing beach on a French Polynesian atoll. The palm tree logs to the right are a Polynesian version of a sea wall. Mathematical modeling of the response of an atoll to sea level rise has assumed that the beach is sandy but this is clearly not always the case.

examine future erosion rates, like the U.S. Army Corps of Engineers' SBEACH model. But it doesn't work! And the widely used models that are based, at least in part, on the Bruun Rule don't work either.

Throughout the technical and popular literature of coastal erosion and sea level rise is a widely reported assumption that for every foot (0.3 m) of sea level rise there will be 100 to 300 feet (30 to 90 m) of shoreline retreat. This number is a product of the Bruun Rule, but few recognize this fact. Unfortunately, it is a misleading assumption. In most cases, a 1-foot (0.3 m) sea level rise will cause much more retreat than the Bruun Rule predicts. Witness the rapid disappearance of the Louisiana coast.

The Bruun Rule equation is the epitome of simplicity. One needs only a navigation chart, an assumption of what the local sea level rise is, and a handheld calculator to predict future shoreline retreat. When the terms in the equation are canceled out, only two variables remain: the rate of sea level rise and the slope of the seafloor adjacent to the beach (the shoreface).

$$R = S \times \frac{1}{\tan \theta}$$

where R is shoreline retreat, S is sea level rise, and θ is the slope of the shoreface.

For a sea level rise of S, the shoreface will shift landward by the amount R, according to the equation. The shoreface typically extends offshore a few kilometers down to a depth of 30 to 50 feet (9 to 15 m), at which point the continental shelf usually flattens out.

It all comes down to an absurdity: the equation assumes that shoreline retreat is some function of the slope of the shoreface, yet no evidence exists that indicates that slope of the shoreface in any way controls shoreline retreat from sea level rise. The general slope of the land surface behind the shoreline is likely to be a much more important control on shoreline movement. Tucked deep within the mathematical model, however, this false relationship between the shoreface and shoreline retreat is safely hidden from the critical eyes of skeptics and nonexperts.

The concept of the rule was the brainchild of the late Per Bruun, a Danish American engineer often considered to be the father of modern coastal engineering. When Bruun first examined sea level rise in the mid-twentieth century, his ideas represented an advance in our understanding of the processes that move sand at the shoreline. For the first time, shoreline erosion was viewed as more than a simple landward retreat of the wet-dry line in the beach sand that marks the contact between the land and the sea. Instead, it was a much larger process that involved the landward retreat of the entire shoreface.

First and foremost, the rule assumes a purely sandy shoreline—no mud, no rock, and not much shell material. Few shorelines really satisfy this assumption. An environmental consultant's efforts to model erosion rates on the Delaware coast made the assumption that the beach and shoreface were entirely sand. Apparently, however, the consulting engineers never ventured away from their computers. The beach in question contains extensive layers of hard, highly compacted and erosion-resistant mud. This material was easily visible.

Similarly, on a West Florida beach, engineering consultants predicting shoreline erosion rates failed to note that the beach was mostly rock, a fact that would have been evident to anyone who had gone to the trouble of wading knee deep into the surf zone. Their projections of shoreline retreat were, like the Delaware example, meaningless.

Australian geologist Peter Cowell used a modification of the Bruun Rule to predict that atoll shores would survive a modest sea level rise by moving up and back just like a sandy beach. The problem is that atoll ocean beaches are mostly rock, not sand. There is very little room for optimism about the future of atoll nations given the current pace of sea level rise. Meaningless projections like this one only provide atoll dwellers with false optimism.

As for predicting sea level rise in general, there is no assurance as to which of many factors affecting shoreline retreat will be most important in coming decades. Nor can one know with certainty how a given factor will play out as a natural process unfolds. No one knows with any specificity, for example, what the intensity, direction, duration, frequency, and timing of storms in coming decades will be. No one knows how much new sand will be supplied in the future to a beach from offshore, from local rivers, from eroding bluffs, and from adjacent beaches.

If the equation is so flawed, why is it still widely used? The most important reason is that the Bruun Rule is the only show in town. No other model purports to predict shoreline retreat from sea level rise. If one wants to come up with a state-of-the-art-prediction, even if it is a poor one, there is no alternative available. We have been told that the newly formed FEMA Climate Change Panel is considering using the Bruun Rule for its prediction of shoreline erosion. Obviously, we think that this would be a huge mistake.

So what is there to do? Can we ignore the problem like the IPCC did in its 2007 report when the panel said it couldn't offer a prediction of the contribution of ice sheets to sea level rise because it couldn't be modeled? If someone asks what the future holds for our shoreline, can we simply shrug our shoulders and say, Who knows? Clearly, we must learn to live with uncertainty and to accept and work with expert qualitative estimates, which themselves still do have much to tell us.

Predictions of shoreline retreat rate are best based not on abstract rules but on observed conditions at a particular shoreline site, including knowledge of the overall sea-level-rise rate. The first estimate of future shoreline retreat is simple projection or extrapolation of the present-day erosion rates. Rates of retreat on most shorelines are due to many things besides sea level change, however. On many (if not most) developed sandy shorelines, the impacts of shoreline structures such as jetties and seawalls and of activities such as channel dredging are currently more important than sea level rise and remain the principal causes of shoreline instability and erosion. Sea level rise can be expected to become progressively more important as a cause of shoreline retreat, but there is no known way to separate sea level rise from other causes of erosion.

Applying geologic common sense to the problem can help to make general predictions of shoreline retreat rates. On glacially scoured rocky coasts like northern Maine, inundation will be relatively important rather than erosion of the hard rock, so slope of the land will be an important factor determining shoreline movement rates. The nature of the rock will be important. Is it easily broken up by waves or easily weathered under local climate conditions? In the Arctic, shoreline retreat rates will depend on how future climate change affects permafrost and the duration of ice-free periods.

For beaches near river mouths, large and small, shifts will depend on how humankind's choices affect sand and sediment supply. For example, the shoreline along the Mekong River mouth in Vietnam is currently building out in a seaward direction. No major dams are on this river (which flows through six countries) as yet, but expectations are that some soon will be. New dams will trap the river's sediment load, and we can expect the shoreline on the Mekong Delta to begin to retreat at a very rapid pace.

On estuarine shorelines (e.g., along the drowned river valleys on the U.S. East and Gulf coasts such as Chesapeake Bay and Galveston Bay), the slope of the land that is being flooded can provide first estimates of likely shoreline retreat rates in the face of sea level rise. The flatter the land, the more rapid the shoreline retreat across it. Open ocean coasts are much more complex, and shoreline retreat is tied up with wave processes that move a lot of sand about.

When considering both the rise of the level of the sea and the retreat of its bordering shorelines and how they both will affect us, we need finally to recognize that there are still a few 800-pound gorillas out there. The largest is the West Antarctic ice sheet, a giant that seems to be stirring. The 2001 United Nations IPCC panel report described the West Antarctic ice sheet as a slumbering giant. In 2006, Chris Ripley, head of the British Antarctic Survey, described the ice sheet as "an awakened giant," and it is to this that we now turn.

Chapter 4

The 800-Pound Gorillas

You never really know your friends
from your enemies until the ice breaks.

Eskimo proverb

*B*Y SOME MEASURES, Bruce Molnia has one of the most interesting jobs around. A glaciologist/geologist with the U.S. Geological Survey (USGS), he has been photographing the terminus of glaciers that flow to the sea in Alaska. The goal of this ongoing study is to compare the location of the terminus today with the oldest available photographs and determine the extent of glacier shrinkage in the intervening years. Molnia began taking these photos on an unofficial basis when he started working as an intern at the USGS in 1968. In 1998, the photography project became official when the secretary of the interior asked the USGS to furnish "unequivocal and unambiguous" evidence of climate change. Survey administrators unanimously agreed that there was no better place to look for it than the Alaskan glaciers. The original old photographs range in date from the late nineteenth century to the early twentieth century, and Molnia has photographed ninety glaciers that terminate in water. (Many of the resulting "before and after" scenes can be found at www.usgs.gov/global_change/glaciers.)

Traveling in vessels ranging from a Zodiac to an old World War

II minesweeper, Molnia used directions or descriptions left behind by the original photographer to locate as best he could the exact spot where an older photo had been taken. The biggest problem proved to be dense, almost impenetrable vegetation that had grown up since the glaciers retreated. The original description might mention a thirty-minute climb from the water's edge that proved to take hours through dense brush. Occasionally, the original photographer left a small rock cairn behind to mark the spot where he stood.

The extent of retreat of the Alaskan glaciers is spectacular. Molnia points out that glaciers retreat most rapidly when they can calve into water. Once they no longer extend into a body of water, they retreat more slowly due only to melting. The Mendenhall Glacier in Juneau, Alaska, for example, has retreated 3.1 miles (5 km) since 1760, but now that a lake has formed in front of it, the glacier is calving and has retreated about a kilometer just since 2000. Elsewhere in Alaska, the Muir Glacier in Glacier Bay has retreated 6.2 miles (10 km) since 1941; the Columbia Glacier in Prince William Sound has retreated 8 miles (13 km) since 1981; and the Guyot Glacier in Icy Bay has retreated 31 miles (50 km) since 1900 and has split into four separate smaller glaciers.

During the twentieth century, most of the state's 100,000 glaciers retreated, thinned, or stagnated, Molnia observes, and of the approximately 650 glaciers large enough to be named, only 20 have advanced in recent years. (This doesn't count the dozens of very small glaciers that continue to add volume at the highest elevations of the Alaska mountain ranges.) Molnia believes that should they all melt, the

Comparison of Alaska's Muir Glacier (top) in 1902 with a photograph (bottom) taken in 2005 from the same location. During the past century, most Alaska glaciers have retreated dramatically; should they all melt completely, their contribution to worldwide sea level rise would be about 2 inches (5 cm). (Edward Burton McDowell. 1902. Bruce Molnia. 2005. Muir Glacier: From the glacier photograph collection. Boulder, Colorado: National Snow and Ice Data Center/World Data Center for Glaciology. Digital media.)

total potential contribution to sea level rise from Alaskan temperate glaciers is 2 inches (5 cm). If all the mountain glaciers of the world were to melt, sea level would rise around 1 foot (0.3 m).

During the Little Ice Age period of cooling (roughly 1350–1850), Alaskan mountain glaciers lengthened and thickened significantly, but as the Little Ice Age ended, a general retreat began. With the beginning of the twentieth century, the ice retreat accelerated. Retreat in the late twentieth and early twenty-first centuries has been very rapid. Similar changes to temperate-zone glaciers have occurred elsewhere in both the Northern and Southern hemispheres. Mountain glacier melting has also been accelerated by a dust cover in some areas. The dust, a product of farming, grazing, and recreational activities (and perhaps global warming), darkens the ice and increases absorption of heat from the sun. Perhaps 20 to 25 percent of the sea level rise during the last century can be attributed to the melting of mountain glaciers. The huge and rapidly melting Patagonia ice fields in Chile and Argentina alone may have furnished 10 percent of the water causing recent global sea level rise.

Mountain glacier ice losses create more problems than just sea level rise. Many societies depend on glacial melt for their water supply for drinking, irrigation, and hydropower, and new sources of water will have to be found for a number of communities located near mountain glaciers. La Paz, Bolivia, and Lima, Peru, are two examples of important cities that depend heavily on glacial melt for drinking water, while India's Kashmir Valley is one example of a large agricultural area dependent upon glaciers for irrigation water. In the Kingdom of Bhutan at the eastern end of the Himalayas, plans are afoot to move some villages because of the expected loss of glacial water supply in the near future. Even Ushuaia, Argentina, the world's southernmost city, may run out of glacier-derived drinking water by 2015 to 2020. In northern Tibet, the melting ice has removed prime pastureland as new lakes form.

For the most part, there is no way to hold back glacial melting. In the Swiss Alps, however, temporary relief has been achieved by covering some small glaciers that ski resorts depend on with plastic sheeting.

The retreat of the mountain glaciers will be an important component of sea level rise at least for the next century or two. But it is the world's ice sheets in Greenland and Antarctica that are truly the 800-pound gorillas in the room of global climate change and sea level rise.

The World's Great Ice Sheets

The world has three ice sheets, vast bodies of ice that tie up more than 75 percent of the world's freshwater. The 1,756,000 mi² (4,550,000 km²) Greenland ice sheet contains 2.8 million cubic kilometers of ice. The Antarctic continent holds an estimated 30 million cubic kilometers of ice divided between two great ice sheets, the East, nearly 4 million square miles (10,400,000 km²), and the West, about 1.4 million square miles (3,600,000 km²) in area. If the Greenland ice sheet totally melted, it would cause sea levels to rise 24 feet (7 m), whereas the two Antarctic ice sheets hold the potential for an astounding additional 187 feet (57 m) of sea level rise.

The thermal expansion of the ocean water due to warming temperatures is more or less a linear event: as temperatures rise, the ocean volume will increase in direct proportion. The sea level rise due to melting ice sheets, however, may be decidedly nonlinear. That is, a one-degree centigrade increase in temperature could cause little change in the melting ice sheet contribution to sea level rise, but another degree or two could cause major, even catastrophic, collapse of the ice sheets that would flood low-lying coastal cities around the world. Although much remains to be learned about the dynamics of the Greenland and Antarctic ice sheets, it is clear that once a certain threshold of ice disintegration has been reached in one of these sheets, melting could accelerate and so would sea level rise. The exact tipping points of such ice disintegration remain conjecture at the present time.

When the United Nations Intergovernmental Panel on Climate

Change (IPCC) published its third assessment report in 2001, Greenland's ice sheet was considered a minor component of likely sea level rise and an Antarctic contribution to sea level rise was not even on the horizon, a matter of millennial concern perhaps, not of decades or centuries. The great southern continent at the time showed no evidence (or at least none was then recognized) of significant rapid change in ice volume. In just the last few years, this perspective has changed. New data regarding ice sheet dynamics suggest that Antarctic ice movement may hold some unpleasant surprises. For example, it was realized that the floating ice shelves at the margins of Antarctica's main glaciers played a role in holding back the flow of ice to the sea. The breakup of those floating ice shelves could thus speed up the rate of flow of glaciers to the sea.

Field studies that are costly to mount and complex to carry out are the mainstay of our understanding of glaciers and ice sheets. It seems that every year, on-site discoveries add a new element or two to our perception of how continent-sized ice masses evolve in a time of global warming. Qualitative mathematical modeling is probably less important at this stage because so much remains to be learned about Earth's ice bodies. You can't model what you don't understand. Instead, satellite measurements have, in the last decade, come to the fore in our observations of large-scale ice sheet evolution.

Three types of satellite measurements are important in studies of the great ice sheets. Each method has its strengths and weaknesses, and for that reason, using more than one approach is always a good idea.

- *Satellite altimetry* is a method of measuring elevation to determine growth or shrinkage of the ice sheets. This has an advantage in determining changes over time because access to remote areas is easy and measurements can be made frequently. The earliest measurements of elevation were done the old-fashioned way, by field surveys conducted by crews wearing very warm clothing. The main shortcoming is that altimeters do not distinguish between soft snow and compacted ice.

- *Satellite gravity measurements* have been used on the world's ice sheets since 2002 with the launching of GRACE satellites. These have the advantage of measuring the mass of the ice by measuring the pull of gravity, thus overcoming the shortcoming of altimeters. If mass is lost, it can be assumed that ice is lost. However, glacial rebound, the uplifting of land surfaces as the weight of ice is reduced, can cause an apparent but not real increase in ice volume.
- *Satellite radar interferometry* measures the velocity of moving ice and also can determine the grounding line position. This information combined with other satellite data has added considerably to our understanding of ice sheet behavior.

Right now, the estimated contribution of the three ice sheets to sea level rise, based on indirect indications from satellite measurements, is relatively small. Losses from both the Antarctic and Greenland ice sheets have contributed from 0.35 to 0.5 millimeter per year of the 3 millimeters per year total sea level rise, according to IPCC estimates; however, a recent study, discussed below in more detail, suggests the total may be twice that. Thermal expansion of the upper layers of the ocean adds 1.6 millimeters per year of sea level rise, and the melting of ice from mountain glaciers contributes 0.77 millimeter per year. These amounts may not sound like much. What makes the situation worrisome, however, is that the volume of water contributed from the ice sheets has recently doubled, and all short-term signs in the field are that the water contribution from the ice sheets will increase significantly in the years to come.

Greenland's Contribution to Sea Level Rise

Greenland, an autonomous province of the Kingdom of Denmark, is the world's largest island and is considered part of North America. Sandwiched between the Atlantic Ocean to the south and the Arctic Ocean to the north, 81 percent of its land area is covered by ice. Mountains line the coasts, but the weight of the ice has depressed the central part of the island as much as 1,000 feet (305 m) below sea level. If the ice were to melt away rapidly, Greenland would likely become an archipelago of islands surrounding a central sea, at least until the continent rebounded from weight-caused depression.

Greenland is about one quarter the size of the United States. Its coastal area was long occupied by Inuits before Vikings from Iceland settled the southwestern coast in the tenth century. Eric the Red was banished from Iceland for some murders and, with a group of followers, settled on Greenland in 984. The Scandinavian settlers remained for five hundred years, corresponding roughly to the Medieval Warm Period of the North Atlantic region. As the Little Ice Age began, starting in the fourteenth century and extending to the midnineteenth century, the Scandinavian settlers vanished. Conditions today at the extreme southwest tip of Greenland are probably about the same as they were when the Norse civilization flourished there. About fifty-seven thousand people live there today and make a living primarily in the fishing industry. There is a significant potential for future petroleum production, and the tourist industry is growing as well, suggesting that the population may grow.

The vast body of ice covering most of Greenland probably first formed at the beginning of the Pleistocene epoch about two million years ago, although it may have largely disappeared several times in the intervening ages. For example, at the time of the last interglacial warm period 120,000 years ago, when the sea level was slightly higher than it is today, Greenland may have been largely devoid of ice. Today the ice sheet extends 700 miles (1,100 km) at its widest point and about 1,500 miles (2,400 km) in length in a north-south direction. Most of the sheet is 1.2 miles (2 km) thick, but in some places it reaches more than 1.9 miles (3 km). Around the sheet's margins are numerous narrow outlet glaciers that slowly move the ice to the sea.

Some of the ice at the base of today's Greenland ice sheet is more than a hundred thousand years old. This old ice provides an important opportunity to study Earth's past. Scientists can extract long ice cores reaching all the way down to the bottom of this ancient ice mass.

Map showing the massive Greenland ice sheet, the total melting of which will produce a sea level rise of 24 feet (7 m). Darkened areas indicate where the ice has been thinning in recent years. Currently, the Greenland ice sheet is producing more meltwater than the West Antarctic ice sheet. The relative importance of these processes is expected to reverse in this century. (Map courtesy of NASA Goddard Space Flight Center)

The ice sheet consists of annual layers, very much like tree rings, of winter snowflakes that have recrystallized into ice crystals. The study of these individual seasonal layers of ice, dust, and gases has provided valuable data on past atmospheric temperatures, carbon dioxide in the atmosphere, volcanic eruptions, forest fires, atmospheric pollution, ocean volumes, solar variability, and desertification.

All glaciers gain ice through snowfall in their upper reaches. They lose ice through calving (when icebergs fall into the sea from the end of the glacier) and through a number of so-called ablation processes. Ablation includes melting, wind and stream erosion, and evaporation of ice (sublimation). The ablation zone is the outer margin of the ice sheet where ice loss exceeds gain. The Greenland ice sheet thickens by about 2.4 inches (6 cm) per year in its central portion. Simultaneously, ablation of surface ice along the sheet's margins and calving of icebergs into the sea both reduce the mass of the ice sheet and add to a rising sea level. The difference between the processes adding ice volume and those subtracting it determines Greenland's sea-level-change contribution.

The flow of ice to the sea during the summer is helped along by huge streams of meltwater that pour down crevasses and through tunnels (moulins) as much as 30 feet (9 m) in diameter. In recent years, the size of these meltwater rivers has dramatically increased. Glaciologists believe that the water is now of sufficient volume to act as a lubricant when it reaches the base of the ice sheet, accelerating its seaward flow and ultimate disintegration at the sea's edge. This was a phenomenon well established in mountain glaciers but was only recently observed in the larger outlet glaciers extending from a main ice sheet.

Ice streams are another phenomenon that hurries ice to the sea. These are bands of ice that flow relatively rapidly within the main body of ice. In 1991, radar interferometry satellites spotted the first known ice stream flowing within the interior ice sheet in northeast Greenland. The Illussat marginal glacier of west Greenland had been receding prior to 2002. Then it began moving in a seaward direction at a rate of 9 miles (14 km) per year, with even higher velocity surges. Some surges have reportedly moved as rapidly as 3.1 miles (5 km) in ninety minutes.

A recently reported crustal "hot spot," where molten rock is close to the surface, may be partially responsible for some of Greenland's ice streams. A hot spot would not be a fundamental cause of accelerating ice loss, but combined with other factors, such as meltwater lubrication and sea level rise, it could play a role.

In a study reported by the BBC, Eric Rignot of NASA's Jet Propulsion Laboratory and Pannir Kanagaratnam of the University of Kansas found that the loss of ice from Greenland doubled between

A summer scene on the Greenland ice sheet showing meltwater pouring down a moulin. This water helps to lubricate the base of glaciers, speeding up the movement of ice toward the sea. In some instances, the glaciers are moving across lakes that are formed beneath them. (© Roger Braithwaite/Peter Arnold, Inc.)

1996 and 2006. This conclusion was based on both gravity and radar interferometry measurements of glacier velocity and glacier thickness. A second study, this one by J. L. Chen and colleagues emanating from the University of Texas and using GRACE satellites, agreed very closely with Rignot's and Kanagaratnam's numbers for the current rate of ice loss. Volume loss in Greenland overall was pegged at about 23 cubic miles (96 km³) in 1996, while in 2006 it was estimated to be almost three times that, 70 cubic miles (290 km³) per year.

This acceleration of ice loss from the so-called outlet glaciers on the edges of the ice sheet was not predicted in IPCC and other mathematical models of Greenland's contribution to sea level rise. Most of this unexpected acceleration of melting was observed in the southernmost glaciers, but it appears that the acceleration trend is moving northward, a probable response to warming, but in any case likely to progressively increase Greenland's contribution to sea level rise.

The Future of Antarctica

The greatest uncertainty in predicting sea level rise over the next century is the future of Antarctica. It is the world's southernmost, coldest, windiest, driest, and highest (on average) continent. The ice sheet, which covers 98 percent of the continent, averages about 1 mile (1.6 km) in thickness and in some places is more than twice that. There is little snowfall except along the coasts and, as a result, most of Antarctica is a desert, the largest on the globe.

The crews of three separate vessels first saw the continent in 1820. On January 28 of that year, Russian explorer Fabian Gottleib von Bellingshausen spotted a shoreline covered with ice in what turned out to be East Antarctica. Three days later, Edward Bransfield, a British naval captain, sighted the Trinity Peninsula (the tip of the Antarctic Peninsula) and noted the presence of rock outcrops, thus assuring that it was land and not an iceberg. On November 17, American sealer Nathaniel Palmer, captain of a 47-foot (14.3 m) sloop, observed the Antarctic Peninsula. The first person to touch the Antarctic mainland may have been another American sealer, John Davis, who briefly rowed ashore on February 7, 1821, to hunt for seals. In 1909, three men from the team of Ernest Shackleton's *Nimrod* expedition became the first to reach the South Magnetic Pole. Two years later, Norwegian Roald Amundsen and four team members were first to arrive at the geographic South Pole.

No humans have ever lived permanently on the continent, but today four thousand scientists from many countries do summer research there and a few hundred overwinter at some of the permanent research stations.

There are several types of coasts in the Antarctic, each of which will react differently to global warming. Forty-four percent of the shoreline consists of floating ice sheets attached to the mainland. Thirty-eight percent are ice walls, that is, glacier margins that are grounded and slowly being pushed seaward while gradually shedding icebergs. Thirteen percent of the shoreline has actively moving ice

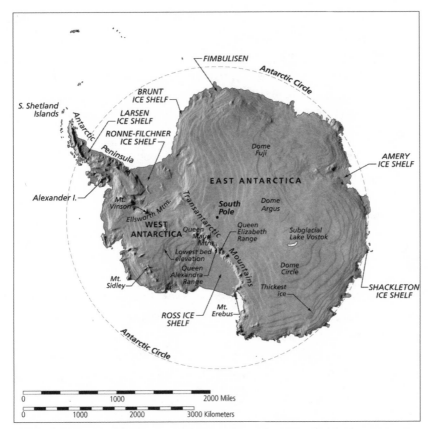

Map of Antarctica showing the two major ice sheets. Currently, only the West Antarctic ice sheet contributes significantly to sea level rise. If it were to melt entirely, sea level would rise 16 feet (5 m).

streams, or outlet glaciers, that are rapidly disintegrating and producing icebergs. Finally, at present a mere five percent of the summer shoreline has bare rock exposed during the Southern Hemisphere summer.

For several centuries, up until about 2000, the total amount of ice in Antarctica was slowly increasing. Recent GRACE satellite gravity measurements, however, show that there is now a net loss in the mass of Antarctica. These ice sheets are so large that losses could occur that have a strong impact on sea level even without catastrophic change on the Antarctic continent itself.

Of the two major ice sheets covering the Antarctic continent, the East Antarctic ice sheet is much larger and older, and is entirely grounded above sea level. Should this sheet melt, it would produce a sea level rise of 164 feet (50 m). Presently, the East Antarctic ice sheet is relatively stable and is not contributing a substantial amount of meltwater to the sea. Recent studies of rock on East Antarctica have indicated that the ice sheet has contracted or expanded much less than the West Antarctic ice sheet during the last few hundreds of thousands of years.

Since the East Antarctic ice sheet is not changing significantly, most of the attention of global change scientists has focused on the smaller West Antarctic ice sheet (WAIS), much of which is aground on the seafloor, sometimes as deep as 1,500 feet (450 m) below sea level. The WAIS is thus mostly a marine ice sheet subject to changes in air and water temperatures and sea level, making the submerged portions inherently more unstable than those above sea level. When a glacier or a portion of the ice sheet becomes ungrounded, it will begin floating again and will quickly move seaward into the zone of rapid melting and calving. At the end of the last ice age, there was a very rapid retreat of the marine-based ice sheets compared with the slower melting of continental ice. This is the basis of the global concern about the possible collapse of this ice sheet.

Instantaneous melting of this entire sheet would produce a global sea level rise of about 16 feet (5 m), significantly less than the rise caused by the disappearance of all of Greenland's ice. The WAIS is a huge mass of ice. If the assumption were made that it would melt in twelve hundred years, it would likely produce a sea level rise of 12 to 20 inches (30 to 50 cm) per century until its melting was complete. If collapse time were only five hundred years, it would contribute 31 to 39 inches (80 to 100 cm) per century.

Broad, flat ice shelves are sometimes attached to the edges of Antarctic ice sheets. These are massive slabs of snow-covered ice, up to

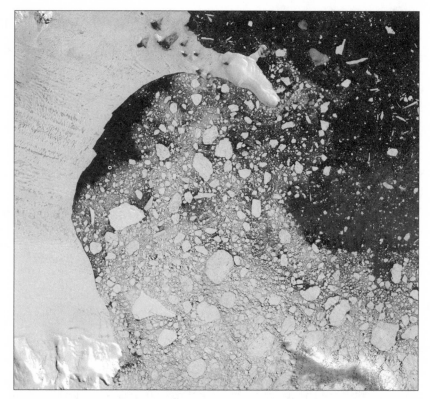

The Larsen B Ice Shelf on the Antarctic Peninsula broke up in thirty-five days in early 2002. Ice shelves form through freezing of seawater and then an annual addition of snow. They hold back the glaciers moving to the sea from the interior of the continent; their loss should accelerate the glacier's flow. A similar loss of the Ross Ice Shelf on the Antarctic mainland could produce a significant jump in the rate of sea level rise. (Image courtesy of LANDSAT 7 Science Team and NASA Earth Observatory, Goddard Space Flight Center)

820 feet (250 m) thick, floating on the ocean surface. Ice shelves gain ice both by snowfall on the upper surface and by freezing of seawater on the underside. There are two such ice shelves on the margins of the WAIS, the Ross Ice Shelf (197,175 mi²; 510,680 km²) and the Ronne-Filchner Ice Shelf (169,850 mi²; 439,910 km²), each nearly the size of Texas. Both are major sources of icebergs that often break or calve

off from the edges but do not contribute significantly to sea level rise because this is ice formed at sea.

The much smaller 1,255-square-mile (3,250 km²) Larsen B Ice Shelf on the Antarctic Peninsula collapsed in 2002. The breakup occurred during January, February, and March, and the event was "captured" by repetitive satellite images, on-the-scene observations from the British research vessel RRS *James Clark Ross,* and flyovers by U.S. and Argentine glaciologists. All told, about 1,000 square miles (2,600 km²) of the ice sheet drifted away in the form of thousands of icebergs, large and small. Once started, the breakup accelerated and much of it occurred in just a few days, between March 5 and March 7. In March 2009, the Wilkins Ice Shelf on the Antarctic Peninsula began to show signs of breaking up. This is happening in the Northern Hemisphere as well. In August 2008, the 20-square-mile (50 km²) Markham Ice Shelf broke up off Ellesmere Island in Canada.

The boundary between grounded ice and floating ice is called the grounding line, and it moves back and forth as the ice sheet expands or contracts. Much of the melting of ice occurs at the grounding line as the grounded ice reaches the sea and begins to float. Melting on the underside of the ice can be particularly rapid as sea temperatures slowly rise. During the ice age maximum, twenty thousand years ago, the grounding line was well seaward of the present line, out to the edge of the continental shelf.

Melting of the West Antarctic ice sheet on the margins of the Amundsen Sea conceivably could lead to collapse of a large portion of the sheet, the lip of which is grounded on a submerged ridge. If the ice retreats, it will back off into deeper water, leading to more "floating" and much more rapid flow and calving at the margins.

Within the grounded portions of the West Antarctic ice sheet, ice flows to the sea in well-defined and relatively fast-moving ice streams. Just as in Greenland, this is the principal means by which ice is brought to the sea. These WAIS streams are 6,500 feet (2000 m) thick and 18 to 50 miles (30 to 80 km) wide, and they extend into the interior 185 to 300 miles (300 to 500 km). The streams can move at a rate of up to 1.2 miles per year (1.9 km/yr). In contrast, the ice sheet areas between the ice streams move at about 16 feet (5 m) per year.

These ice streams are essentially faster-moving glaciers within glaciers. Often they may have water-saturated sediment at their base. Like the Greenland glaciers, the Antarctic ice streams may be helped along by water flowing along bedrock surfaces, and, in some cases, the ice may actually float on glacial lakes found at the boundary between ice and the bedrock. The velocities of these ice streams vary with time, and each ice stream exhibits its own unique pattern of flow. Their future pattern of behavior is critical to the future stability of the WAIS. If these ice streams remain active and grow in length, extending inland farther toward the center of the continent, the rate of ice loss will only increase.

In Antarctica, outlet glaciers are also affected by the high tides that occur with each springtime full moon. These can cause the sea-wardmost few kilometers of a moving glacier to float upward, thus reducing the friction with the ground surface and speeding up its flow for a few hours. This has not been observed in Greenland as yet but is most likely another mechanism that hastens calving and melting.

An international panel of twenty-five European and American polar ice experts met in Austin, Texas, in March 2007 at the West Antarctic Links to Sea Level Estimation (WALSE) workshop. In a March 28 news release, the group concluded that both the ice shelves and the grounded ice sheets of the WAIS are thinning, some at a "surprisingly rapid" rate. There is a possibility, the panel noted, that at some point "runaway melting" of the grounded ice sheets could occur, but this is a process that is not understood and cannot be predicted. The experts argued that changes in wind patterns, which they attribute to global warming, had led to increased upwelling of relatively warm waters (from the water mass known as the Circumpolar Deep Water) near the margins of the ice sheet. This warm water, in turn, increases melting of the edge of ice shelves and causes the glaciers to flow more rapidly. The group estimated that in 2006, global sea levels were raised a full half-millimeter by the Antarctic ice loss. This is the same amount as the estimates for sea level rise caused by melting of Greenland ice, a view that is now widely accepted.

These are very short-term observations, and it remains to be seen whether these patterns of continued high rates of melting will continue, even increase. Most observers agree that these changes in melting patterns are due to global climate change, and if that is indeed the case, short-term melting rates are likely to become long term.

There is an interesting feedback relationship between Greenland and Antarctica in a time of melting. As one or the other or both melt, the rate of ice loss of the various ice sheets will increase. Sea level rise in effect lifts the ice sheets in the same fashion as spring tides have been observed to do. Lifting of the ice masses, no matter how slight, results in increased rates of calving and of ice flow to the sea.

In a 2008 *Nature Geoscience* report, Eric Rignot and colleagues concluded that Antarctica is gaining in the middle but losing much more at the edges. They compared losses from 1996, 2000, and 2006. Losses from Antarctica were estimated at 112 billion tons for 1996 (equivalent to 0.3 mm/yr sea level rise) compared with 196 billion tons in 2006 (0.5 mm/yr rise), an increase of more than 75 percent. These numbers were derived by comparing measured losses from the ice sheet (using satellite radar interferometry) with mathematically modeled annual snow volumes (the weakest aspect of the data).

The surprise 2006 observation by University of Colorado scientists Isabella Velicogna and John Wahr (based on GRACE satellite gravity data) that the Antarctica ice cover is losing as much as 36 cubic miles of ice per year (150 km³)supports Rignot's results. Based on computer modeling, the 1991 third IPCC report predicted that rather than lose ice, the Antarctic would begin to gain ice, for a while at least, as snowfall increased due to warming of the ocean water. Clearly, the models were wrong.

Is the loss of Greenland or Antarctic ice going to increase and create a catastrophe this century? In our view, a 3.3-foot (1 m) sea level rise in this century would be a disaster, a 7-foot (2 m) sea level rise would be a catastrophe, and a higher rise is not out of the realm of possibility. Tad Pfeffer and colleagues in a September 2008 *Science* magazine article looked at possible sea-level-rise scenarios based on climate models and past history of ice changes. They concluded that the maximum amount of sea level rise this century would not exceed 7 feet (2 m). David Vaughn of the British Antarctic Survey argues that collapse of the West Antarctic ice sheet is a real possibility. But he and his colleagues also note that "it is not clear whether current changes are simply normal decadal variations in the ice sheet or are the beginning of a significant deglaciation." A 2009 report in *Science Daily* reviews an article from *Climate Change* journal, written by Aslak Grinsted and colleagues from the University of Copenhagen, that argues that the sea level rise in this century will be at least 3.3 feet (1 m).

James Hansen, a NASA climatologist, strongly believes that a dramatic sea level rise is very much in the realm of possibility. Hansen says that "an ice sheet response time of centuries seems probable and we cannot rule out large changes in decadal time-scales, once wide-scale surface melt is underway."

The endpoint here is that we are all left in limbo by the uncertainties of the various ice behavior models and field and satellite observations on these vast ice masses. It is clear that most predictions of the future of sea level rise are higher than the IPCC's rather nebulous estimates. It appears that both the West Antarctic and Greenland ice sheets are undergoing major changes that should lead to a significant acceleration in the rate of sea level rise. Indications are that this may be a long-term phenomenon rather than a passing anomaly, but there is much need to increase the effort of field and satellite observations of the world ice bodies to chart the nature of these changes.

For planning purposes, we think that a 7-foot (2 m) sea level rise by the year 2100 should be assumed. It is thus our belief that coastal management and planning should be carried out assuming that the ice sheet disintegration will continue and accelerate. This is a cautious and conservative approach.

~ Chapter 5 ~

A Sea of Denial

. . . all eyes and no sight.

WILLIAM SHAKESPEARE, *Troilus and Cressida*, act 1, scene 2

THOUSANDS OF tide gauge records, some extending back more than a century, show an expanding ocean. Years of sea level measurements by satellite indicate the same trajectory. Diverse forms of field evidence such as rims of dead trees with drowned roots along coastal lagoons, the ubiquitous thinning by erosion of coastal plain barrier islands, intrusion of salt water into coastal aquifers, and the backing up of drainage and sewage systems of coastal cities all point directly to rising seas in modern times. Furthermore, the astounding retreat of the world's mountain glaciers since the mid-twentieth century and the more recent—and more important— indications of degradation of the world's great ice sheets both point to continuing and likely accelerating sea level rise.

Nevertheless, there remains a noisy minority of opinion that sea-level-rise concerns have been way overblown. This argument derives from a variety of groups, almost entirely nonscientists, including conservative think tanks, certain political conservatives (especially conservative talk radio hosts), Libertarians, and special-interest groups, mainly coal and oil companies, who feel threatened by societal concerns over global change.

One particular lightning rod for criticism has been Al Gore's documentary *An Inconvenient Truth*. When it was announced that the 2007 Nobel Peace Prize had been awarded jointly to Al Gore and the twenty-five hundred scientists participating in the United Nations Intergovernmental Panel on Climate Change (IPCC), the chorus of criticism intensified. Fox News morning commentators prefaced announcement of the prize with the question, "What does Al Gore have in common with Yasser Arafat and crazy Carter [former president Jimmy Carter]?" And things went downhill from there.

With respect to the future level of the sea, the furor over Gore's documentary derived from the statement that the disintegration of portions of the Greenland and Antarctic ice sheets could result in a 20-foot (6 m) sea level rise. This fact is unquestionably true; however, several skeptics manipulated the remark to suggest that Gore actually contradicted the IPCC by predicting a 20-foot sea level rise for the next century, which he didn't.

What's the Beef?

The opposition to the current scientific thinking about sea level rise, its magnitude, and its origins has taken a variety of forms. In the sampling below, deniers are categorized, partly with tongue in cheek, as to the nature of their opposition to current scientific understanding of sea level rise.

THE DENIALISTS

Nils-Axel Mörner, University of Stockholm: Sea level rise is "the greatest lie ever told."

Robert Felix, author of *Not by Fire but by Ice:* Sea level is falling in the Atlantic (and the Indian, Pacific, and Arctic).

THE MISLEADERS

Glen Beck, Fox News commentator and author of *An Inconvenient Book:* Gore assumes a rise of 240 inches (610 cm), which is seventeen times the IPCC midrange estimate of 15 inches (38 cm). *Beck compares apples and oranges by contrasting the Gore statement about the sea level rise that would occur with a contribution from the ice sheets, while the IPCC specifically excludes Greenland and West Antarctic ice sheet melting.*

Patrick Michaels, Cato Foundation: Al Gore's film exaggerates the rise by about 2,000 percent. *More apples and oranges.*

THE BELITTLERS

Christopher Monckton, former policy adviser to Margaret Thatcher: In a recent Keynote address to the 2009 International Conference on Climate Change hosted by the Heartland Institute, Monkton scoffed:

> Where are they all today, those bed-wetting moaning minnies of the apocalyptic traffic-light tendency—those Greens too yellow to admit they're really Reds?

followed by,

> Now, if we're going to exaggerate, let's exaggerate properly. Sea level is going to rise not by Gore's 20 feet, not by Hansen's 246 feet, but by 2,640 feet. Half a mile. You heard it here first. There goes Andy Revkin of the *New York Times*, dashing to the telephone to tell them to hold the front page.

THE AGNOSTICS

Richard S. Lindzen, Massachusetts Institute of Technology: "The Greenland ice sheet is actually growing, on average, which creates the pressure that is forcing glaciers to flow seaward." *Lindzen's point is that the Greenland ice sheet is not contributing to sea level rise, which is almost certainly wrong.*

S. Fred Singer, University of Virginia: The West Antarctic ice sheet will continue to melt (barring a new ice age) and should be melted away in six thousand years. *The point is, by Singer's reckoning, sea level rise will continue at an extremely gradual rate. In other words, sea level rise is a millennial problem for society and not an immediate worry for this generation.*

Arthur Robinson, Oregon Institute of Science and Medicine: Sea level would be expected to rise about 1 foot (0.3 m) during the next two hundred years. *This expectation is considerably less than the sea level rise rate of today even leaving aside probable increased contributions from Greenland and Antarctica.*

Jim Martin, director of the Colorado Department of Public Health and Environment, noted, "You could have a convention of all the scientists who dispute climate change in a relatively small phone booth." Although it would take more than a phone booth to enclose the authentic researchers who are skeptics, the number of genuine earth or atmospheric research scientists with advanced degrees who deny the human role in global warming is indeed small.

Many of the prominent skeptics are nonscientists or specialists in some other field. Of course, economists, sociologists, political scientists, and many other specialists must be involved in providing guidance for societal decisions of how to respond to sea level rise. However, understanding the future of the ice sheets, the future of atmospheric temperatures, the impact of CO_2 emission reduction, the rates of sea level rise, and the impact of warming on the biosphere are subjects that belong to the realm of natural scientists.

Naomi Oreskes, science historian from the University of California at San Diego, conducted a study of the scientific consensus on global change and the response of the media to the issue. In more than nine hundred technical papers on global climate change, she found none that did not accept, implicitly or explicitly, at least a partial human role in global change. But in examination of media reports and popular literature, roughly 50 percent of the articles question a human connection to global warming.

The media itself is often the problem. There is a strong inclination on the part of journalists to focus on controversy and the contrasting views of dueling experts even when there is little disagreement within the scientific community. This journalistic practice has fed a false public impression that the strength of the case for global climate change is weak.

Professor Nils-Axel Mörner, a prominent skeptic of impending sea level rise and the author of a pamphlet entitled "Sea Level Rise, The Greatest Lie Ever Told." (Photo courtesy of Nils-Axel Mörner)

The Endpoints

The wide range of opinion on the future of sea level rise can be conveniently sandwiched between two individuals, both well-known scientists. Nils-Axel Mörner, a Swedish geologist, has staked out the most extreme endpoint of all: complete denial of sea level rise. James Hansen, an American climatologist, has proven to be the most able and respected proponent of the strong likelihood of a damaging sea level rise.

The Nils-Axel Mörner Endpoint

Of the strong global change critics, Mörner has one of the more impressive backgrounds. He was once head of the Department of

Paleogeophysics and Geodynamics of the University of Stockholm; from 1999 to 2003, he was president of the International Union for Quaternary Research (INQUA) Commission on Sea Level Changes and Coastal Evolution; and he has published widely on various mechanisms of sea level rise. A colorful character, he is also an advocate of dowsing (water witching) and has been challenged to prove his abilities by psychic "fraud buster" James Randi—an offer Mörner declined.

Mörner's principal beliefs include the following:

- "Sea level is by no means in a rising mode and we can free the world from the condemnation to become flooded in the near future. In about 40 years we will be in a new solar minimum, and hence are likely to experience a new little ice age."

- "If you go around the globe, you find no [sea level] rise any-where." *Mörner believes sea level rise exists only in computer models and doesn't exist in the real world, which denies mounds of field evidence to the contrary.*

- The sea level isn't rising around the Tuvalu atolls. The saltwater intrusion into fresh groundwater cited as evidence of sea level rise has happened because Japanese pineapple farms drained the island, allowing salt water, aided by recent typhoons, to intrude under the island. *Saltwater intrusion is occurring on a number of atolls without pineapple farms, and the threat from sea level rise to the world's atoll nations is very real.*

- The sea level is dropping, not rising, around the Maldives in the Indian Ocean. The sea level drop is due to evaporation. *Evaporation can be a local cause of sea level change as can storms, wind pattern or precipitation changes, coral reef die-off, and many other events, but all indications are that the sea is rising here. Ironically, the Maldives, as discussed elsewhere, is the nation that may lead the world in preparation for sea level rise.*

- Tide gauge data are undependable. *Tide gauges are sometimes unreliable measures of true sea level rise due to ocean volume changes because they can't distinguish changes caused by movement of the land and changes due to eustatic sea level rise. But from the hundreds of gauges in use, evidence is overwhelming that the sea is rising.*

- Sea-level-rise data from satellites have been manipulated (falsified) by the IPCC to show a rise where there is none. *To come up with a final measurement, satellite sea-level-rise data must be massaged—not for purposes of falsification, but in order to account for atmospheric conditions. Falsification of all the data would entail a conspiracy among many scientists of mind-boggling proportions, and there's been no reason to doubt the honesty of the numerous scientists involved in satellite studies.*

- Sea level changes are more closely related to heating caused by solar radiation than to CO_2 content of the atmosphere. Today we are at the peak of the current sunspot cycle, and we are in more danger of cooling with a sea level drop than warming. *Variations in solar radiation, among other factors discussed in chapter 2, do affect global climate change but are not the primary cause of present-day sea level rise. Total solar output has varied in recent decades by 0.1 percent, not enough to have an important impact on today's global climate change.*

- Twenty-two individuals, none of whom were sea-level-rise experts, authored the 2001 sea-level-rise IPCC report. *Not true.*

The James Hansen Endpoint

James Hansen, perhaps the nation's top climatologist, is a strong believer in mathematical models but also someone who listens closely to the earth. He is director of NASA's Goddard Institute for Space Studies and adjunct professor of Earth and Environmental Sciences at Columbia University. Hansen is a quiet yet extraordinarily outspoken scientist who urges other scientists to come out of their laboratories and into public view. He is that rare combination of a cautious scientist who speaks to the public.

Hansen's public standing jumped big-time in 2006, thanks to twenty-four-year-old George Deutsch, a politically appointed NASA public affairs official. Deutsch had rejected media requests to interview

*Dr. James E. Hansen, a lead-
ing climatologist with NASA,
obtained particular prominence
during the George W. Bush
administration when his super-
iors tried to prevent him from
giving public speeches about
global change. (Photo courtesy
of James Hansen)*

Hansen and ordered administrators to replace Hansen at public lec-
tures he had been invited to present. Deutsch was forced to resign
when it was discovered that he had manufactured a nonexistent Texas
A&M college degree in his resume. It was one of many instances that
have come to light of George W. Bush administration attempts to
throttle science and the bad news of global warming. In this case, the
move backfired for the administration: Hansen overnight became an
even greater media celebrity, able to present his message of warning
about sea level rise in highly visible fashion.

Hansen faults the IPCC for its reticence and failure to alert the
world in stronger language about the potential disaster that lurks in
the behavior of the ice sheets of the Antarctic and Greenland. In a
2007 *Science* report, Hansen looked at past projections of the IPCC
and concluded that the panel's CO_2 estimates have been right on

target, while temperatures actually increased somewhat faster than projections, and sea level rise increased notably faster than prior IPCC estimates.

According to a 2007 interview with *The New Scientist*, Hansen believes, as did Albert Einstein, that speaking out politically at key moments is part of a scientist's responsibility. He also rejects the idea that scientists should pose as completely objective fact machines that refrain from offering opinions that aren't purely scientific in nature (even about subjects that they know better than anyone else). "There's a big gap between what is understood by the scientists at the forefront of the research, and what is known by the people who need to know," he says. "And that's partly because of this technical language, and limitations on what scientists are willing to say" out of cautiousness or concern about losing funding.

Of all the points the IPCC considers, Hansen believes sea level rise is the most important for our society to address because of the inertia of nature. That is, once ice sheets start to go, it is questionable whether they can be stopped. In his opinion, 350 parts per million (ppm) CO_2 content in the atmosphere is the tipping point for sea level rise. Since the current CO_2 content is already 385 ppm and rising, Hansen argues that global reduction in CO_2 production is essential. (We would argue that the number of model uncertainties involved in obtaining the 350 ppm number as a possible tipping point are too large to assume the number is accurate. What the models probably can say with reasonable assurance is that there is a tipping point between 300 and 400 ppm.)

Hansen is critical of the IPCC on several other counts. He believes the temperature projections the panel came up with would produce more disintegration of the ice sheets than the IPCC implies. Most of all, Hansen is concerned about the business-as-usual attitude of the scientific community—fallout from scientific reticence. That is, one IPCC report on sea-level-rise projections took the Antarctic ice loss and half of the Greenland ice loss for the decade 1993–2003 and assumed that this rate of loss of ice would continue to 2100.

Assuming such linearity in melt rates doesn't make sense, according to Hansen. Already the rate of post-2003 melting of the great ice sheets has increased.

So, at one endpoint Mörner completely denies that sea level rise is occurring, and at the other endpoint Hansen believes that sea level rise will be well above the IPCC's projection. Numbers as high as 15 feet (4.6 m) by 2100 pass by Hansen's lips, not as a prediction but as a possibility that is not inconceivable and should not be ignored.

We concur with Hansen's views. Global warming minimizers and deniers are almost certainly wrong. Many uncertainties remain about sea level rise, the most important of which is the behavior of the world's ice sheets. Hansen's cautious warning that we may be beyond the tipping point of these giant ice masses should be a call to action for all who are concerned with the rising seas.

Sea Level Rise and Polar Bears

A strange type of criticism of global climate change science emanates from the University of Pennsylvania's Wharton School. According to professor of marketing Scott Armstrong, of 650 references in a chapter of the 2007 IPCC report, not one used what he terms evidence-based forecasting methods. Because various research papers don't cite the articles on forecasting he thinks they should or follow the criteria he advocates, he concludes, "We have been unable to find a single scientific forecast to support global warming" and "to date we are unaware of any forecasts of sea levels that adhere to proper (scientific) forecasting methodology."

Armstrong is the editor of and main inspiration for *Principles of Forecasting*, a book with forty contributors who are purported experts in forecasting but none of whom are natural scientists. The book was intended to provide guidance for forecasting in the fields of economics, sociology, and psychology. The principles of evidence-based methods, or scientific forecasting procedures, number a huge and ponderous 140 items that one must largely satisfy to make an acceptable forecast in Armstrong's view. These include such things as "obtain decision-maker agreement on methods," "combine forecasts from approaches that differ," and "test assumptions for validity." Many of the principles are valid for the fields for which they were intended, but to apply them to natural science and global warming predictive studies is simply unworkable if not unjustifiable.

We certainly believe that criticism of the IPCC mathematical models is appropriate. In fact, as we discuss in chapter 3, we are skeptical of quantitative mathematical models on a broad front. Natural science forecasters, however, operate on an entirely different playing field than the Wharton School forecasting approach advocated by Armstrong.

Armstrong's judgment of scientific research on the basis of his principles, which few scientists have ever heard of, is both somewhat arrogant and dangerous. It is arrogant because it is akin to criticizing a Christian because she does not practice Jewish traditions or like blaming a chef for doing a lousy job as an auto mechanic. It is dangerous because noncompliance with Armstrong's chosen principles could be used to trash not only most modern climate and sea-level-rise science but much of natural science in a fashion not at all understood by the nonscientific public.

Indeed, this has already happened. The State of Alaska, which wants to prevent listing of polar bears as endangered (because of potential negative effects on Arctic development), hired Armstrong and associates to analyze research by the U.S. Geological Survey (USGS) that favored listing the bears. The problem is that polar bears depend on sea ice for their existence as a feeding platform to obtain seals, their principal food. The USGS, based on projections of Arctic sea ice loss in the future, concluded that two-thirds of the polar bears may be lost by 2060.

According to an analysis by Armstrong and his colleagues, in one important USGS bear study 72 of the 140 principles Armstrong advocates apparently were violated. It was a puzzling conclusion because, for the most part, compliance or noncompliance with the 140 fore-casting principles he refers to cannot be determined with any certainty by simply reading a technical paper.

In news releases, based largely on Armstrong's observations, the state characterized polar bear science as "unsupported conjecture." But in May 2008, a reluctant Dirk Kempthorne, secretary of the interior under George W. Bush, announced that the bears had been put on the list of threatened species. It is believed to be the first species so listed because of the effect of global warming. Kempthorne took care in his announcement to assure Alaskans that this should not have much impact on future development of America's Arctic region.

We think Armstrong is missing the point. The fact that forecasting global warming is difficult does not mean that global warming is not happening. As we have indicated earlier, there is ample field evidence that the planet is warming.

In the future, we can expect to see a long list of reports from the Wharton School attacking all manner of carefully constructed climate science and sea level studies. The potential damage is huge, and the scientific community will not know what hit it until the fallacy of the Armstrong harmless-sounding, "evidence-based" scientific forecasting tactics is more widely recognized.

Manufactured Doubt

Industry is also biasing the debate about sea level rise. It is to the advantage of coal and oil interests to keep the oil pumps and coal mines and profits rolling along. Solid evidence of industry ties with anti-global-warming organizations was furnished in 2006 by the British Royal Society, in a letter to ExxonMobil complaining of its funding of climate change studies. According to the society's own surveys, ExxonMobil furnished $2.9 million to thirty-nine groups that have misrepresented climate change science by denying the evidence.

In a report entitled *Smoke, Mirrors and Hot Air: How Exxon-*

Mobil Uses Big Tobacco's Tactics to Manufacture Uncertainty on Climate Change, the Union of Concerned Scientists made a similar claim in the United States arguing that ExxonMobil funneled $16 million into forty-three advocacy organizations between 1998 and 2005. Major benefactors of the ExxonMobil program included the American Council for Capital Formation, the American Enterprise Institute, the American Legislative Exchange Council, the Cato Institute, the Competitive Enterprise Institute, the Frontiers of Freedom Institute, the George C. Marshall Institute, the Heritage Foundation, the National Center for Policy Analysis, and the Weidenbaum Center of Washington University.

Perhaps most insidious of all is ExxonMobil's funding of the Property and Environment Research Center. Holly Fretwell, a research associate of the center, wrote a 2007 children's book entitled *The Sky's Not Falling: Why It's OK to Chill Out About Global Warming.* The book contends (on pages 33–36) that the melting of the world's ice sheets won't have much effect on global sea levels—as if 200 feet (610 m) weren't "much effect"! This is the single most inaccurate description of sea level change past, present, and future that we have seen anywhere in any publication.

David Michaels, a professor of environmental and occupational health at George Washington University and former department of energy assistant secretary for the environment in the Clinton administration, points out that vilification of research that in some way threatens an industry has become commonplace in parts of corporate America. It is a process called manufactured uncertainty or manufactured doubt.

Manufactured uncertainty takes advantage of the fact that uncertainty is an inherent part of good science. Scripps Institution climate expert Richard Somerville notes, "In all but the most trivial cases, science does not produce logically indisputable proofs about the natural world. At best it produces a robust consensus based on a process of inquiry that allows for continued scrutiny, re-examination, and revision."

In *Shaping Science with Rhetoric,* Leah Ceccarelli calls the debate "manufactroversy," short for manufactured controversy. She argues that "it is perverse to continue debating an issue that has already been settled for the vast majority of scientists merely so that policymakers will delay action." Those who manufacture the controversy about global change and sea level rise argue that this is an important debate and that they are simply following the tenets of good science. After all, it is important in science to be skeptical, to recognize that there are two sides to every story and to support open debate. The problem, of course, is that it is difficult to confront those who claim these values as their own without appearing unscientific. "The manufactured controversy must be exposed for what it is—the assertion of an important scientific debate where none exists," Ceccarelli suggests.

Steve Milloy is among a number of individuals who are professional uncertainty manufacturers. Milloy's web site JunkScience. com offers a $125,000 reward to anyone who can prove in a scientific manner that humans cause catastrophic climate change. Of course, the type of proof (either to show that humans cause current climate change or, for that matter, that they don't) that Milloy would require (he will be the judge of the validity of the science) is not possible.

The "junk science" correspondent for Fox News, Milloy has worked with Philip Morris, R. J. Reynolds Tobacco, and ExxonMobil and has founded several nonprofits that are critical of environmental science. Milloy has provided "expert opinions" in favor of the use of asbestos, DDT, and certain FDA-banned foods and in opposition to the provisions of the Clean Air Act and to the greening of General Electric. In reality, junk science turns out to be simply science that distresses the captains of industry.

Manufactured uncertainty is a disaster for public understanding of actual climate science. Some of the legitimate criticisms and concerns about details of global warming science are stilled, and the public, understandably unable to distinguish completely between manufactured and real science, is left in limbo; meanwhile, those politicians and policy makers dedicated to nonresponse find support for inaction.

The source of funding for those who promote doubt is often unknown. The Oregon Institute of Science and Medicine in Cave Junction, Oregon, had a single full-time, paid faculty member, Arthur Robinson, and an income of $1 million in 2005, according to Source Watch. Where its funds come from is unclear. The institute garnered a lot of fame for a misleading 1998 petition (the "Oregon Petition") sent to thousands of scientists. The petition was accompanied by what was falsely claimed to be a reprint of a scientific paper published in the *Proceedings of the National Academy of Sciences*. The paper, "Environmental Effects of Increased Atmospheric Carbon Dioxide" by Robinson and others, argued that the negative effects of global warming were overblown and asked readers to sign a petition of support.

Eventually, Robinson claimed nineteen thousand petition signatures, but when pressed on the issue, admitted that almost none of them were climate researchers; since Robinson did not furnish addresses and affiliations of signers, the validity of the whole exercise was cast further in doubt. Among the signers were the fictitious characters Drs. B. J. Honeycutt, Benjamin Pierce, and Frank Burns of the TV show *M*A*S*H*.

In 2008, a reprint of a new paper with the identical title, again authored by Robinson and others, appeared, along with an anti-global-warming petition, in the mailboxes of scientists around the country. The reprint is from the 2007 *Journal of American Physicians and Surgeons,* an impressive sounding journal title but in reality an outlet for political advocacy of a Libertarian bent. The paper argues that increased CO_2 is leading to more robust plant growth and that reduction of CO_2 output from fossil fuel consumption will be harmful to the third world. Robinson argues that there is a 150-year-long trend of sea level rise averaging 7 inches (18 cm) per century. Since the rise predates the rise in hydrocarbon use, he argues, the two must be unrelated. Assuming past patterns of alternate rising and leveling of the sea level curve, Robinson suggests a 1-foot (0.3 m) rise in the next two hundred years is to be expected, an entirely unrealistic estimate not based on any known observations or models.

In what we believe may be the most meaningful poll of all, Peter Doran and Maggie Zimmerman published the results of a survey of 3,146 of their fellow earth scientists in the January 9, 2009, *EOS, Transactions of the American Geophysical Union*. The names were taken from the Directory of Geoscience Departments of the American Geological Institute. In this carefully controlled poll, 82 percent of the scientists believed that human activity played a role in global warming. Broken down into specialties, the results showed that only 47 percent of petroleum geologists agreed but that 97 percent of active climatologists accepted the human role.

What do polls by Time, CNN, Fox News, Gallup, and other groups tell us about public recognition of climate change and sea level rise? They seem to be very consistent with regard to public opinion. About 85 percent of Americans agree that global warming is upon us, and about 70 percent believe that humans are responsible for at least a part of the warming. That doesn't sound like manufactured uncertainty has been a huge success. Another number, however, indicates that the manufactured doubt industry—or at least how their spokespersons are covered by the media—have had some success: about 65 percent of Americans believe that there is a lot of disagreement among scientists on global warming and its significance. The fact is that while there is much disagreement among scientists on details, there is little disagreement on the big picture of global warming.

Sorting through Claims

Googling almost any aspect of global warming, including sea level rise, provides a blizzard of opinions stretching from the negative to the positive, from the cynical to the bizarre, from the well-documented to the anecdotal. How can a nonscientist separate the wheat from the chaff? Here are a few suggestions.

Avoid the Manufacturers of Doubt. Why give credence to those who do not seek the truth? Those who for one reason or another oppose the current scientific view of global change and who generally minimize sea level rise include, among others, Heartland Institute, George C. Marshall Institute, American Legislative Exchange Council, Frontiers of Freedom, Competitive Enterprise Institute, American Enterprise Institute, Cato Institute, Property and Environmental Research Center, Heritage Foundation, and Center for the Study of CO_2 and Global Change. These and other assemblages of deniers can usually be identified by their energy-related (oil and coal companies) or Libertarian sponsorship. Sometimes their statements have an element of truth and occasionally they speak the whole truth, but we recommend that nonscientists give their prolific works little credibility.

Ignore Declarations from Nonscientists. Recently, a prominent astronaut was quoted as saying that global cooling was about to commence because of changes in solar radiation. Maybe that is so and maybe not, but a pilot, no matter how skilled, is not necessarily an expert on climate change. Even though we have defended Al Gore's numbers in this book, we would prefer you turn to scientific sources like realclimate.org or sciencedaily.com that directly reflect the findings of climate researchers.

Beware Long-Term Conclusions Based on Short-Term Events. Evidence that global change is occurring is necessarily long-term—decades long, at a minimum. A single storm or a warm winter (or two) doesn't mean anything. The spectacular increased degradation of the Antarctic and Greenland ice sheets since only 2004 (and their potential for sea level rise) is impressive, but continuation for another five years will make it more so.

Be Cautious with Mathematical Model Results. Much of the evidence for predictions of future trends in global change is based on mathematical models. Taken qualitatively, they provide a solid case for global change—quantitatively is another matter. Field data are the most reliable measures of global change and present a compelling case for concern.

Listen to Researchers, but Use Common Sense. Give highest credibility to scientists reporting on their own research. Two particularly credible science journals that often report on sea level issues (and are often quoted in the media) are *Science* and *Nature*.

The alert reader will perceive that we have recommended that the nonscientist follow a path of assuming guilt by association in order to learn what scientists think about global warming. That is true enough and regrettably so. But we don't know how else the interested and intellectually curious nonscientist can otherwise succeed in an effort to learn the scientific view. The manufacturers of doubt, even when they are providing an accurate vision of sea-level-rise science, put a special and often invisible spin on it.

We have recommended instead that nonspecialists listen only to those involved in active research on global change. Whether a source of information is a researcher or not isn't always easy to determine, and it's important to recognize that holding a degree in a specialty does not make one an active researcher in that subject.

Ultimately, we have faith in scientists to tell the truth. The naysayers would like you to believe that climate scientists are just another interest group with an inherent bias to speak in unison and defend an ideology. This couldn't be further from the truth. Getting scientists to agree is like herding cats. If a young climate scientist could find definitive proof that CO_2 was not responsible for global warming, he or she would love nothing more than to overthrow the old guard and enjoy fifteen minutes of scientific fame. One does not build a reputation as a scientist by stepping in line, but by standing apart from the crowd. Scientists are too absorbed in their own work to host a global conspiracy. So, our final advice is to let the scientists ferret out the truth.

— Chapter 6 —

The Living Coasts

During those long ages of geologic time, the sea has ebbed and
flowed over the great Atlantic coastal plain. It has crept toward the
distant Appalachians, paused for a time, then slowly receded, sometimes far into its basin; and on each such advance it has rained down
its sediments and left the fossils of its creatures over that vast and
level plain. And so the particular place of its stand today is of little
moment in the history of the earth or in the nature of the beach—
a hundred feet higher, or a hundred feet lower, the seas would still
rise and fall unhurried over shining flats of sand, as they do today.

RACHEL CARSON, *The Edge of the Sea*

RACHEL CARSON was a prescient observer of many things
biological and environmental beyond her most famous
warnings on the effects of DDT and other pesticides in
her 1962 book, *Silent Spring*. In her 1955 book, *The
Sea Around Us,* for example, she recognized the role that changing
sea level played in altering coastal environments and the biological
communities that inhabit those coasts. The book did not directly
warn of problems associated with the modern rise in ocean level, but
it is interesting that scientists and naturalists like Carson understood
the ephemeral nature of shoreline positions more than a half century
ago, some far longer.

When the shoreline moves, the coastal ecosystems associated with that shoreline must also move—or perish. Coastal wetlands— marshes and mangrove forests, for example—and coral reefs are biological communities that are directly dependent on the position of sea level. That these ecosystems, which represent some of the most productive and biologically diverse environments on our planet, have survived two million years of sea level fluctuation is a testament to their robustness and adaptability. Throughout the Quaternary period (the last 2.5 million years), these plants and animals have responded to global changes in sea level by wandering landward and seaward, and to changes in climate by shifting northward and southward. In fact, the modern distribution of coastal ecosystems is a direct reflection of this give and take with the changing physical environment. So, what's the problem?

It's us, as mentioned earlier. Understanding the future of coastal wetlands and coral reefs in light of the modern expansion of the world's oceans involves an examination of how human activities may hinder (or in isolated cases enhance) the ability of coastal ecosystems to respond to what may be great changes in their physical environment. The problems that both corals and wetlands will have to cope with in order to survive into the next century of rising sea level include the following:

- Direct destruction of these habitats through dredging, harvesting, or human residential or industrial use
- Indirect destruction, in which development or human activity is blocking their avenue of expansion
- Pollution and changes in sediment input
- Changes in seawater chemistry, salinity, and temperature

Geologists have a long-held axiom that guides the way they explore and understand change on our dynamic planet. This is the principle of uniformitarianism, which can be summarized by the phrase "the present is the key to the past." Uniformitarianism was championed by Charles Lyell, the man many deem to be the "father of geology" and the author of what is arguably the first book defining the field. The present and past are inexorably linked, explained Lyell.

For today's geologist, the reverse of the uniformitarian's creed is also true: "the past is the key to the present and the future." The best way to understand, or even predict, the future of any geological or biological environment is to examine how that system has responded to change in the past. This is essentially the approach to much global change research.

In today's world, there is a fundamental problem with this approach, however. It is true that as long as there have been oceans, coastal systems have migrated back and forth along with changing sea level. But this cycle has run into the "arrow" of the evolution of human society, which has ever been on the increase. Never before have coastal wetlands and coral reef ecosystems had to respond to changes in ocean volume and attendant shoreline movement *and,* simultaneously, to massive changes in the physical environment wrought by human activities.

In light of these facts, it is understandable that there would be considerable uncertainty and disagreement as to how coastal biological communities will fare during the next century. They are facing a future that is truly unprecedented.

Coastal Wetlands

Wetlands, whether along fresh- or saltwater bodies, exist at places where water meets the land. They share characteristics of both the aquatic and the terrestrial environments. They typically have standing water present at least occasionally, and they have emergent vegetation—plants sticking up out of the water.

The extent of wetlands in any particular coastal setting is determined largely by the slope of the adjacent land. Where the slope is steep and the terrain rocky (as along much of the west coast of the Americas), the wetlands are typically small and narrow. But where the coastal zone is flat and broad, such as in the Florida Everglades, the vast wetlands of the Mississippi River Delta in coastal Louisiana, and the Sundarban mangroves of the Ganges Delta of India/Bangladesh, coastal wetlands may be extensive.

Although there is a variety of coastal wetland types, coastal marshes and mangroves dominate many of the world's coastal areas. Both are clearly affected by sea level change, and both have experienced tremendous decline in the last hundred years as a result of human activities. Salt marshes and mangrove forests often are biologically extremely productive and fuel both terrestrial and aquatic food webs. In addition, such wetlands contribute to estuarine water quality by removing excess nutrients and pollutants originating from the mainland. Wetlands furthermore provide protection against coastal erosion and offer important habitat for crabs, shrimp, fish, birds, and a wide variety of organisms. Many estuarine species of commercial importance rely on these habitats for successful completion of their life cycles.

Marshes

Ask someone to think of a coastal wetland, and many will picture the quintessential tidal salt marsh. Tidal marshes exist at mid-to-high latitudes along shorelines worldwide. They may reach tens of miles inland if the tidal range is large enough (as along the U.S. coast in Georgia), and they are dominated by grasses, sedges, and rushes.

Salt marsh ecosystems tend to have limited plant diversity, simply because living in even slightly salty water is very stressful for a plant. Very few can tolerate the salt; even those marsh plants that can do so require special adaptations. Salt marsh cordgrass (*Spartina alterni-flora*), a plant that requires daily inundation by salt water, can actually excrete salt crystals onto the edges of its leaves. Other common marsh plants, the glassworts (*Salicornia* sp.), which are quite delicious in salads, sequester salt in their fleshy, upper stems. Eventually, those stems drop off, eliminating the salt.

Although salinity may moderate the diversity of salt marsh vegetation, it does not eliminate its biological productivity. Of all ecosystems, salt marshes have one of the highest rates of primary productivity as measured by the amount of photosynthesis and the amount of organic matter they produce. Tidal marshes benefit from a continuous influx of nutrients with each tidal cycle. It is the perfect setup for growing grass.

This high productivity yields an ecosystem with numerous valuable functions. Salt marshes are important habitat for a wide variety of species, many of them of commercial importance. Furthermore, they trap pollutants and supply nutrients to surrounding waters. Unfortunately, they are in trouble worldwide.

Human activities have already destroyed more than half of the world's wetlands. Although it is difficult to quantify the global loss of coastal salt marshes, data are available for some regions. Canada, for example, has lost approximately 65 percent of its tidal and salt marshes, while the state of California has lost a staggering 91 percent of its coastal wetlands. These losses stem from a variety of uses, including drainage for agriculture, clearing for shrimp aquaculture, filling and draining for human settlement, and dredging for ports and harbors.

The good news for coastal marshes is that they are well attuned to respond to rising sea level. Because the marsh surface can rise upward on the order of centimeters per year by natural accumulation of organic and inorganic sediment, coastal marshes can increase their elevation as sea level rises. In addition, coastal wetlands, if given the opportunity, may creep landward as sea level rises, converting upland to wetland. Like barrier islands, coastal wetlands can maintain themselves by changing their position on the landscape.

Marshes *can* migrate landward in response to rising sea level, but the big question is, will they? Sadly, the overwhelming answer to that question in most parts of the world is a resounding NO. In many areas, natural barriers limit the potential for wetland expansion. Wetlands along rocky or mountainous coasts, for example, typically have little space for expansion because the upland slope is so steep. Marshes in these areas will require even greater care as they attempt natural expansion. But the biggest global obstacle to marsh expansion is shoreline development and agriculture. As a species, we have

Masonboro Island, North Carolina, is migrating toward the mainland into the extensive Spartina salt marshes behind the island at a rate of approximately 16 feet (5 m) per year. Barrier island migration occurs when storms cause the ocean shoreline to erode back while simultaneously widening the backside of the island by washing sand into the marsh behind.

occupied so many areas that fringe coastal marshes that there is little room for them to expand.

Many estuarine shorelines are ringed with seawalls and bulkheads; the marshes in front of these structures will simply disappear as sea level rises. In the United States, construction of second homes along these estuarine marsh shorelines skyrocketed over the last decade as oceanfront property became less and less affordable. The new "house at the beach" is actually on the estuarine shore.

With the demise of these coastal salt marshes, we will lose clean water, fisheries, habitat, and beauty. The losses will be both environmental and economic. Unfortunately, few governments recognize this crisis, and even fewer are doing anything about it.

Some nongovernmental groups are raising the alarm, however. The National Wildlife Federation, for example, commissioned a series of regional reports examining the impact of rising sea level on salt marshes. All of these reports identify making upland expansion of these marshes possible as the most critical strategy for enhancing chances for marsh survival in a regime of rising sea level (along with halting the outright destruction of coastal marshes). UNESCO makes the same argument for managing coastal wetlands in numerous reports released over the last two years. Without more concerted action, however, the future does not look good for the world's salt marshes. Marshes will survive, but they will occupy a much smaller area; thus, the benefits that they provide society will be greatly reduced.

Mangroves

The word *mangrove* is a catch-all term for a diverse group of more than a hundred salt-tolerating plant species, thirty-five of which are considered to be true mangroves. The term is used to refer to individual plants as well as to the community. Mangroves are the primary coastal wetland plant community in warm, tropical waters and are found in a broad band circling the globe, roughly between latitudes 25° north and 25° south. Mangrove patches—referred to as mangrove swamps, mangrove forests, or mangals—develop in areas of low waves, which is why they are frequently found on shorelines protected by coral reefs. Mangals line approximately 60 to 70 percent of the world's tropical ocean shorelines and can be found along the coasts of Asia, Africa, Australia, and the Americas. The total global areal extent of mangrove forests is 66,400 square miles (172,000 km²), 15 percent of which is in Brazil, the most in any country.

In a mangrove forest, individual species typically occur in zones or bands parallel to the shoreline. The location of each species is controlled by differing tolerance to such factors as salinity, frequency of tidal flooding, and wave energy. Florida has three species of mangroves, the red, the black, and the white. The red mangrove (*Rhizophora mangle*) can grow as tall as 80 feet (24.4 m) with the classic tangled roots and 6-inch (15 cm) pencil-shaped propagules ("baby trees").

A mangrove forest lining the shoreline of Maputo Bay, Mozambique. These mangrove forests are inundated at almost every high tide and have an uncertain future because of both logging activities and sea level rise.

In Florida, this species occupies the seawardmost position along the edge of the mangal. The black mangrove (*Avicennia germinans*) grows at a slightly higher elevation than the red mangrove and under ideal conditions can grow as high as 65 feet (20 m). The root systems are long and horizontal with vertical pneumatophores, projections used in respiration that protrude out of the water column. The white mangrove (*Laguncularia racemosa*) can grow to 50 feet (15.3 m) in height, occupies the highest ground, and has no visible roots.

In the Southern Hemisphere, where more than a dozen species may coexist in a single mangal, each individual species will live close to its tolerance limit of salinity. Minor changes in sea level or sedimentation rates could result in rapid killing of one species and/or replacement by another. Thus, mangrove species distribution will be a sensitive indicator of sea level rise and other coastal changes.

As the oceans warm, the range of mangroves is likely to expand both to the north and to the south as new colonies become established from floating propagules. A demonstration of this type of shift can be observed in the coastal waters between Miami Beach and Cape

Canaveral along Florida's east coast. Today, mangroves are widespread along this coastal reach, but prior to the arrival of Europeans the "lagoons" were freshwater lakes. Salt water replaced freshwater as inlets were cut from the ocean into these lagoons for shipping, and mangroves quickly made their way up to Cape Canaveral, mostly in the twentieth century.

Mangroves face the same challenges that salt marshes face. They are experiencing direct and indirect destruction as a result of human activities. As with salt marshes, mangroves also have the ability to migrate landward in response to rising sea level, but this pathway is often blocked by coastal development; consequently, the area of mangrove wetland will simply shrink and disappear as the oceans rise. According to the Mangrove Action Project (MAP), an environmental group with a global interest in mangrove protection and research, roughly half of the world's mangroves have been lost in recent decades.

Like salt marshes, mangroves have many other functions that are valuable to society. They are critical nurseries for a huge number of marine species, including most of the commercial fishes found in the nearshore zone. Mangroves are also home to thousands of species of birds, insects, mammals, reptiles, fish, crustaceans, and invertebrates. They protect fragile populations of charismatic species, including tigers in Bangladesh and jaguars in Brazil and Colombia, as they seek their prey at low tide.

Until recently, mangrove forests had been considered by many societies and governments to be useless swamps. It is easy to see why. Mangroves are impenetrable tangles of branches and roots; they are home to snakes, spiders, saltwater crocodiles, crabs, and other creatures that humans fear. They may pose a true health concern by harboring mosquitoes carrying malaria and dengue fever. In fact, researchers track the probable location of *anopheles* mosquitoes, the malaria vector, by mapping mangals.

As a consequence, mangrove destruction has often been government supported and has even been viewed as a method of improving

A tangled mass of mangrove roots on the lagoon side of a Pacific barrier island near Buenaventura, Colombia. Like salt marshes, mangrove forests are the basis for an important ecosystem consisting of various plants and animals that take advantage of the protection afforded by the root systems as well as the nutrients supplied to the nearshore waters. Mangrove forests also serve as a buffer against the onslaught of storms.

public health. Removal of the mangal can endanger public health in other ways, however. The existence of mangroves around Homestead, Florida, for example, protected the community from direct wave attack by Hurricane Andrew in 1995. The Ranong area of Thailand was relatively unaffected by the massive Asian tsunami of December 26, 2004, because of the protection afforded by a healthy band of mangroves at the shoreline. And the death toll from that tsunami would likely have been considerably lower if widespread removal of mangroves and sand mining from beaches had not occurred throughout the whole affected region.

So far, humans have done much more damage to mangals than sea level rise has. In East Africa, mangrove wood is used for boat construction. Just like the live oak trees used for ship building in the

United States during the eighteenth and nineteenth centuries, the original shape of the branches was maintained for the keel and other timbers. Dredging, clearing for agriculture and aquaculture, diking, oil spills, herbicide applications, and development have all significantly reduced mangrove areas globally.

Of these, clearing for agriculture and aquaculture is currently the greatest threat to mangrove ecosystems. From Ecuador to Vietnam, aquaculture (primarily for shrimp) is a relatively new and fast-growing threat. As much as 50 percent of the mangrove destruction in recent years has been for shrimp farms. Shrimp farming involves clearing the mangrove forest and constructing diked ponds in its place. The problem for the local ecosystem doesn't end there. In this typically large-scale monoculture, when diseases appear (as they inevitably do), the industry usually ends up abandoning the site and clears away another patch of mangroves at a new location. Recolonization of abandoned cleared areas is often slow because the substrate is polluted and inappropriate for mangrove growth.

In many cases, it is poverty in the developing world leveraged by greed in the developed world that drives coastal ecosystem destruction. Julio Cruz lives in the small town of Yusquare in southern Honduras and works for an international company that has removed significant portions of mangrove forest along the Golfo de Fonseca for shrimp farming. He is troubled by the destruction, but Don Julio needs the work and has a family to feed. He wants to send his children to school. He is also troubled by the temerity of the Greenpeace boat that visits the area occasionally to shame the Honduran government for neglecting to protect this ecosystem.

As with salt marshes, mangroves are flexible and dynamic ecosystems capable of adjusting to and surviving sea level rise. After all, mangals have survived huge changes in the level of the sea during the last two million years. Because of their efficient method of seed dispersal, there is little possibility that individual mangrove plant species will disappear. But as the story of Julio Cruz may suggest, we must first protect mangroves from ourselves.

An example of the devastating impacts that aquaculture is having on mangrove forests; in this case, mangroves have been cleared for shrimp farming in Honduras. Other examples of extensive mangrove removal for shrimp farming are found in Equador and Vietnam. (Greenpeace/Francesco Cabras)

Coral

There is something about the peaceful, colorful, underwater world of the coral reef that captures the popular imagination. No other ecosystem provides in crystal clear water the dazzling colors along with stunning biodiversity and unusual creatures. Millions of people worldwide enjoy recreational scuba diving and snorkeling over reefs. Many more bring coral reef fragments into their homes to adorn saltwater aquariums.

Corals are found just about everywhere in the oceans—from the tropics to the poles, and from the shallow zone of breaking waves to true depths of 20,000 feet (6,100 m) on the ocean basin floor. But *hermatypic* corals, the ones that actually build reefs, inhabit only warm (68° to 82° F; 20° to 28° C), shallow waters. The future of nonreef, or *ahermatypic,* corals is probably a secure one. They live in a wide range of environments that are unlikely to be particularly affected by sea level rise. And if they are stressed in one environment, their free-floating larval stages should enable them to reestablish in a new one. On the other hand, reef corals and the huge assemblage of organisms that make up a coral reef are today far more vulnerable and much less mobile than they once were. By some (pessimistic) estimates, 70 percent of the world's coral reefs may be gone by mid-century due to direct human impacts and changing climate.

More so than coastal wetlands, coral reefs are tourist destinations in themselves, but their socioeconomic importance extends far beyond that: the economies of many communities and even some countries are dependent on coral reefs. They are the foundation of many local fishing industries; there are, for example, thirty species of commercial reef fish around the Caribbean alone. Reefs provide a physical barrier, a kind of offshore breakwater, protecting thousands of miles of island and mainland shorelines from open ocean waves. Virtually thousands of tiny harbors, sheltering fleets of small fishing vessels, owe their existence to coral reefs.

Viewed from the air, coral reefs typically appear as a white line formed by breaking ocean waves. Barrier reefs, just like their sandy counterparts, barrier islands, are discontinuous ridges, separated by gaps or channels. Many a desperate shipwreck survivor in the movies has made it through a narrow gap in the line of waves to crawl ashore on a deserted island. The line of surf crashing on the crests of coral reefs not only makes for a panoramic movie shot, but plays a major role in reef development. Nutrients and food for the entire reef ecosystem are distributed about, sediment that might settle on the coral heads is brushed away, and the water is kept well oxygenated.

In 1943, Marine Corps landing craft went aground on the fringing reef surrounding the island of Tarawa, forcing the marines to abandon their crafts and wade ashore through murderous fire from the Japanese defenders. Intelligence operatives of the time had disastrously missed the telltale white line of offshore breaking waves.

Globally, up to three million species of marine organisms live in, on, or very near the world's coral reefs. Reefs are particularly important spawning and breeding areas. Twenty-five percent of all marine species are found there, making reefs the most biologically diverse environments in the modern ocean. The loss of these delicate and fragile ecosystems would be a biological disaster of global proportions.

Tropical coral reefs exist in a broad band circling the globe between latitudes 30° north and 30° south. They are found in clear, warm water in depths of less than 150 feet (45.7 m). All hermatypic reef corals require shallow and well-lit water, normal ocean salinity, and a relatively low amount of nutrients in the water column. The water must be clear to allow a unique symbiotic relationship between the coral animal (or polyp) and a microscopic plant, a dinoflagellate called zooxanthella. Zooxanthellae live in the skin of the coral polyp. Polyp colonies, in turn, live within the skeleton that makes up the reef structure. Coral may be the only life-form that is overtly animal, vegetable, and mineral!

As plants, zooxanthellae require light for photosynthesis and can't exist in water too cloudy or too deep for sunlight to penetrate. Through photosynthesis, the zooxanthellae produce oxygen and nutrients for the polyps. The polyps return the favor by furnishing carbon dioxide for use by the zooxanthellae, which helps the individual corals to precipitate a calcareous skeleton to surround and protect them. Because coral reefs need sunlight, rising sea level will cause deeper corals to be abandoned to the darkness, and the reefs must migrate into shallower water or grow upward in order to survive.

The upward growth of different species of coral varies considerably. In the Caribbean, plate-shaped stony corals (e.g., genus *Agaricia*) are slowest, massive or head corals (e.g., genus *Montastrea*) are a bit faster, and the fastest growers of all are the branching elkhorn corals (*Acropora palmata*) and staghorn corals (*Acropora cervicornis*). According to measurements by Smithsonian geologist Ian Macintyre in the Caribbean, head coral reefs grow upward at 0.3 feet (0.1 m) per 100 years; elkhorn coral reefs can keep up with a sea level rise of 1.6 feet

(0.5 m) per 100 years, and staghorn coral reefs can grow upward of 4 feet (1.2 m) per 100 years. Thus, healthy elkhorn coral reefs are capable of keeping up with what is the widely observed sea-level-rise rate in low to midlatitudes, and healthy staghorn corals should have no problem as well. Of course, healthy is the keyword. A healthy unstressed coral reef can respond much more effectively to environmental changes than can a reef in trouble.

Ian Macintyre and fellow geologist Conrad Neumann from the University of North Carolina recognize three modes of coral reef response to sea level rise: keep-up reefs, catch-up reefs, and give-up reefs. *Keep-up reefs* are those that grow rapidly enough to keep up with sea level rise. *Catch-up reefs* are those that were killed by earlier sea level rise but then have reestablished themselves at a more landward location. *Give-up reefs* grow slowly and will be drowned by sea level rise. Most of the living coral reefs that we see today are of the keep-up variety, having kept up with the modest sea level rise of the last few thousand years. On many deep slopes off reef-lined shores are submerged ridges, give-up reefs, some as deep as 300 feet (91 m), that were overwhelmed by rising sea levels at the end of the last ice age. In the short term, sea level rise might actually benefit many reefs. Corals on reef tops would have a renewed life, whereas during the last few thousand years they have largely been limited to horizontal growth by a relatively stable sea level.

The impact of sea level rise on coral reefs does not occur in isolation. Numerous events and processes work simultaneously with sea level rise to affect the survivability of coral reefs. There are a number of natural diseases that affect coral reefs, for example, the susceptibility to which is increased by the stresses that humans place on these reefs. Humans drive ships that crash into reefs, dredge channels through them, drag anchors over them, and spill oil on them. Humans also mine reefs for building blocks and build artificial beaches on adjacent shorelines that cause suffocating sand to flow out to the reefs. An estimated three thousand tons of living coral chunks are shipped annually to waiting home aquarium owners.

Coral reefs also suffer from increased turbidity. A cloudy water column blocks light, prevents photosynthesis, and kills the coral. This can happen very quickly. In 1995, the island of Roatan, which lies off the Caribbean coast of Honduras, had one of the most pristine reefs in the developed Caribbean. One could wade into knee-deep water on West Bay beach to see a beautiful patch reef, but now the beach is a sad reminder of how quickly we can "love a place to death." By 2005, that reef was decimated, its demise directly related to the island's "discovery" by tourists. Construction activities flooded the reef with sediment and nutrients, reducing visibility and encouraging algal growth.

A few decades ago, the beach at Waikiki, Hawaii, had almost disappeared due to seawall construction. The beach loss loomed as a disaster for the tourist trade, so sand was imported by freighter from Los Angeles and spread out on the beach. Nature continued the offshore spreading, and soon sand killed portions of the nearshore reef. The resulting gaps in the reef allowed storm waves to penetrate to the beaches, increasing the rate of erosion and necessitating repeated installation of artificial beaches with no end in sight. Hal Wanless, geologist at the University of Miami, has shown that mud from beach nourishment projects is killing patches of corals along the east coast of Florida as well. Wanless believes that turbidity from Florida beach nourishment projects may prevent coral migration to the north as the seas warm.

Some argue that the most worrisome environmental problem for the future of coral reefs will be ocean warming (rather than rising sea level) from global climate change. The frequency of high-temperature episodes is expected to increase as mean ocean temperatures increase. The warming of ocean temperatures can lead to the loss of zooxanthellae and death for many corals, a disease called coral bleaching. Coral reefs can withstand higher than normal temperatures for about a month, but if the high temperatures extend for two months, the corals die. In addition, the increased carbon dioxide in the atmosphere is also increasing the acidity of ocean water, a process termed ocean acidification, weakening coral skeletons.

   As is so often the case, studies of the past offer important insights into the present and the future. Paleontologist J. M. Pandolfi of the University of Queensland and his associates studied coral reefs that were preserved in sea cliffs that had been uplifted by mountain-building forces on Papua, New Guinea. Clambering up the cliffs, they collected samples that indicated massive coral reef die-offs every fifteen hundred years or so between eleven thousand and four thousand years ago. Some of the die-offs were the result of the reefs being buried in volcanic ash, but others were due to unknown causes, perhaps bleaching or other disease.

   The good news revealed in the cliff face was that the reefs always recovered from their catastrophic losses. The bad news is the comparison with modern coral reefs. Today reef die-offs are occurring all around the world at a rate at least ten times that of the reefs preserved in the cliffs on New Guinea.

   Natural catastrophes can also take their toll. The December 26, 2004, earthquake that produced the tsunami in Southeast Asia that killed many thousands of people also killed many miles of fringing reefs. On western Sumatra and on the Andaman and Nicobar islands in the eastern Indian Ocean, coral reefs were uplifted out of the water as much as 10 feet (3 m), causing their instantaneous demise.

   Clearly, many coral reefs have recovered from previous disasters. What about the future? There is a vigorous debate ongoing among experts regarding the future of the world's coral reefs. There are so many stressors and so much uncertainty that prediction is difficult. A too-rapid rise in sea level will certainly drown reefs. What is too rapid? Will coral reefs be able to expand their range poleward as the oceans warm? What will happen as ocean water chemistry changes in response to changing climate and atmospheric composition?

Coastal wetlands and coral reefs are dynamic systems that have survived sea level rise in the past by moving landward or seaward, north or south. In the future, coastal wetlands and coral reef ecosystems must meet the challenge of sea level rise alongside the direct impacts of human development and other activities. Currently, direct destruction of wetlands and reefs by humans is probably a greater threat than rising sea levels, although this could change. And certainly, if we are ever able to stem the tide of ecosystem destruction, we face the sad realization that our towns, cities, and developments leave no room for the future migration of the world's wetlands and coral reefs. The future may be bleak for the living coast unless we begin planning to accommodate its movement and expansion. And when our coastal cities and towns become threatened by rising sea level, will we give these natural ecosystems high enough priority to assure their survival?

— Chapter 7 —

People and the Rising Sea

Tuvalu and Global Warming

I hear the waves on our island shore
They sound much louder than they did before
A rising swell flecked with foam
Threatens the existence of our island home.

A strong wind blows in from a distant place
The palm trees bend like never before
Our crops are lost to the rising sea
And water covers our humble floor.

Our people are leaving for a distant shore
And soon Tuvalu may be no more
Holding on to the things they know are true
Tuvalu my Tuvalu, I cry for you.

And as our people are forced to roam
To another land to call their home
And as you go to that place so new
Take a little piece of Tuvalu with you.

Tuvalu culture is rare and unique
And holds a message we all should seek
Hold our culture way up high
And our beloved Tuvalu will never die.

Dame Jane Resture, *a native of the atoll nation of Kiribati*

*P*EOPLE HAVE had to deal with changing sea levels for tens of thousands of years. Some of the earliest humans must have lived next to the rapidly retreating shorelines that followed the end of the last ice age more than ten thousand years ago—a period when sea level was sometimes rising more than 7 feet (2 m) in a century. Society then was not encumbered with beachfront condominiums and vast coastal cities at low elevations. Now the time has come for today's populations to deal with the huge impacts of an expanding ocean, but it will be much more difficult for us to move back. Sea level rise will have an impact on land use of every sort, on parks, natural areas, subways, communications, sewers, roads, railroads, and buildings by the thousands. Whole towns, even nations, will disappear. A refugee problem of enormous magnitude is likely.

Sea Level Rise and the Ancients

A 2008 National Public Radio program entitled "Rising Sea Levels of Alexandria" brought into focus the long history of sea level rise that has left ancient parts of the Egyptian city submerged well offshore from the seawalls that protect today's Alexandria. The city sits on the edge of the Nile Delta, a site Alexander the Great chose two thousand years ago for its excellent potential as a centrally located harbor. The harbor entrance was guarded by the Pharos Lighthouse, one of the seven wonders of the ancient world, the ruins of which were discovered on the Mediterranean seafloor in 1994.

Alexandria is one of several submerged ancient cities along the shores of the Mediterranean, all victims of some combination of global ocean expansion, land subsidence, and relative sea level rise caused by local tectonic forces and storms. The evidence suggests that the Mediterranean Sea simultaneously engulfed the two cities Menouthis and Herakleion, along with Alexandria, sometime around 740 AD (the age of the youngest coins found in the rubble). The fact that the pillars and statues in all three cities largely lie in the same orientation on the seafloor is an indication that an earthquake destroyed them. The fact that the pillars and statues exist at all is evidence of catastrophic submergence of the city during an earthquake. A surf zone cannot slowly move across the remains of a city during a slow sea level rise without pulverizing the ruins.

A famed submerged city of the New World is Port Royal, Jamaica, founded in 1654 and submerged in a 1692 earthquake. By the 1670s, wild, freewheeling, and diverse Port Royal was a larger and more important merchant port than the staid, puritanical city of Boston, with a character about it that was decidedly missing from Boston. In the 1692 tremblor, most of the storehouse and port facility district fell into the harbor, ultimately killing perhaps five thousand people, well more than half the town's population.

Instantaneous sea level rise (and sometimes sea level drop) caused by earthquakes is a surprisingly widespread phenomenon. In addition to the examples of sunken ancient Mediterranean cities, there are many more recent instances of tectonic sea level change. Hundreds of miles of Alaskan shoreline south of Anchorage, for example, suddenly dropped from 1 to 4 feet (0.3 to 1.2 m) in the 1964 Good Friday earthquake, and along Colombia's Pacific coast, 50-mile-long (80 km) reaches of shoreline drop 2 or 3 feet (0.6 or 0.9 m) with every small earthquake on the local continental shelf. During the Great Tumaco Earthquake in 1979, the resulting tsunami killed two hundred and fifty people (almost all the residents) in the remote Colombian fishing village of San Juan de la Costa. It was a double whammy. Simultaneously with the earthquake, the barrier island sank, causing sea level to rise instantaneously by perhaps 1 meter (3.3 feet) and the erosion rate to accelerate, destroying most of the village's remaining buildings within a few years.

Sea Level Rise in America's Past

Some American towns and villages, too, have gone to sea in centuries past. These seaside villages existed in a time when people who lived along hazardous shoreline reaches did not expect government help.

Many communities are threatened with the same fate today, but we have more resources to pour into the sea, for better or worse, to try to hold shorelines in place using seawalls and other measures. The most important lesson from these past episodes, as we'll see in the cases that follow, is that storms are the active agent of sea-level-rise destruction, and just as in the past, they are almost always the immediate precipitant of disaster. In our modern era of rising sea level, storms will become more important as their impacts push farther into the interior of the beach communities, across barrier islands, and into coastal cities.

Last Island (Isle Derniere), Louisiana (1856)

Ocean-facing barrier islands make up most of America's shoreline along the eastern and Gulf coasts, extending in an almost continuous chain from Long Island's southern shore to the Mexican border. In the nineteenth century, Last Island was a retreat for the wealthy and privileged of Louisiana because, among other things, it provided a place to escape from the yellow fever epidemic that hit New Orleans in 1853. According to the U.S. Geological Survey, the 20-mile-long (32 km) Last Island was, at the time, a single barrier island with a mature maritime forest. The average elevation of the island was then probably about 5 feet (1.5 m) or less, with both significant shoreline erosion and relative sea-level-rise problems due mainly to subsidence, or sinking of the delta.

In 1856, there were around a hundred homes on Last Island. In the Last Island Hurricane of August 10 of that year, a storm surge destroyed a multistory hotel and submerged the town's other buildings, killing at least 190 of the 400 vacationers there. Many of the bodies were carried more than 6 miles (10 km) inland of the island. Those who survived had clung to pieces of wreckage and to a grounded schooner that had arrived too late to carry them off the island. Hurricane parties seem to be an essential part of the legend of all hurricane disasters, and the Last Island Storm is no exception: a raucous

A scene of Waveland, Mississippi, after Hurricane Katrina. Devastation was almost total for the first three blocks of this low-lying coastal community. A number of buildings destroyed in this 2005 storm were replacements for buildings previously destroyed thirty-six years earlier in Hurricane Camille. When do we abandon such vulnerable communities? (Photo by Andy Coburn)

party was said to have been under way at the hotel when the storm swarmed ashore. A single cow was the only animal to survive.

Storms are, in a sense, often the advance guard of sea level rise; in the absence of such a large storm, Last Island might have remained habitable as a resort, but only for a time. Since the Last Island Storm, Isle Derniere has experienced a 3.3-foot (1 m) sea level rise due to a combination of the expanding ocean and delta subsidence and has been cut into many smaller islands. The hurricanes of the last two decades have put the finishing touches on the island chain, turning them into mostly submerged sandbars, although there have been efforts to maintain them through beach nourishment.

A Faith Cornish Murray sketch of pre–Civil War Edingsville, South Carolina, showing fully clothed vacationers promenading on the beach. This village consisted of sixty buildings, including two churches, a general store, and a billiard parlor. It met its final demise in the Great Sea Island Hurricane in 1893. Today, Edingsville Beach is a narrow barrier island just tens of feet wide and the remnants of the community are on the seafloor one-quarter mile from the shoreline. (Photo courtesy of Charles Spencer and the History Press; first published in the Charleston News and Courier, 1955)

Edingsville Beach, South Carolina (1893)

Edingsville Beach was one of the earliest upscale seaside resorts in the South. A playground for rich planters from the mainland and from adjacent Edisto Island (where Sea Island cotton of the highest quality was grown), it was the *in* place for the upper crust of Charleston, 70 miles (113 km) away. For residents of South Carolina's Low Country, a house by the sea offered some relief from the oppressive summer heat and the malaria-bearing mosquitoes. Edingsville Beach was first settled in the early nineteenth century and grew to a village of sixty homes, two churches (Presbyterian and Episcopalian), a summer school for boys, and in 1852 the Atlantic Hotel, built by the Eddings family. Many of the houses were two-story brick mansions with grand stairways leading up to the front porch.

Life on the island in the summer was that of the old aristocratic South with regattas, horse races, parties, and banquets, but it all came to a halt with the Civil War. Although the island remained untouched by the marauding armies (it was hardly a target of strategic importance), the wealth of the aristocracy that supported Edingsville Beach was lost in the maelstrom of war.

Growing concern over the impact of rising sea level on the South Carolina Low Country has resulted in a very unique billboard campaign along Interstate 95. Government response to sea level rise will require public education to gain support for long-term action.

Then came the hurricanes. Beginning in 1881, a series of storms struck the island, removing dunes and damaging houses. These culminated on August 28 with the Great Storm of 1893, otherwise known as the Sea Island Hurricane, followed on October 13 by a category 3 storm that struck the South Carolina coast. Edingsville was no more. Most of the village just disappeared, and the site it once occupied is now well out on the continental shelf.

The island that supported that community is now a narrow barrier island, 30 to 50 feet (9 to 15 m) wide, rapidly migrating toward the mainland. At low tide, the beach is now wider than the above-sea-level island! The buildings of Edingsville were left behind by the landward-moving shoreline. Bricks and household relics reside in abundance on the adjacent continental shelf and still appear on the beach after storms.

A final note from the state of South Carolina: the Low Country around Charleston was once full of colonial-era rice plantations, most of which were abandoned after the Civil War. Rice is a freshwater wetland plant; thus, the swampy coastal regions of South Carolina were perfect for the crop. A walk around one of those abandoned rice plantations with Ernie Wiggers, director of the Nemours Wildlife Foundation, provided an interesting lesson in the inland reach of rising sea level. According to Wiggers, rice can no longer be grown on many of these former plantations because salt water has crept too far inland (reduced precipitation can also play a role in this). This is an example of how salinization may cause problems for agriculture or aquaculture worldwide.

Diamond City, North Carolina (1899)

One of the largest of the early towns on the Outer Banks of North Carolina was Diamond City, located in what is now Cape Lookout National Seashore.

New England whalers began showing up on the Cape Lookout whaling grounds as early as the 1720s, about the same time local fishermen began spending more and more time on the island. Initially, the Outer Bankers occupied seasonal "shacks" offering shelter from the weather, but eventually they built permanent homes and remained on the island year-round. In due course, the village had a school, stores, and church services and may have consisted of five hundred souls at its peak. No tourist spot this—Diamond City was a working community.

Occasionally, the whalers anchored their vessels in the shelter of the cape, and when the crews spotted whales, rather than hauling anchors and setting the sails on the mother vessel, they jumped into their whaling skiffs, rowed out, and harpooned them.

This whaling technique was not lost upon the locals, and after the shipboard whalers abandoned the Cape Lookout whaling grounds for greener pastures, the locals began their own whaling industry. Lookouts atop the high dunes spotted the whales and sounded the alarm. In a good year, five whales would be taken, hauled to the beach, and rendered, a process that could take two weeks and involved most members of the community. When the whales weren't running, the community fished for mullet and porpoise.

In the waning years of the nineteenth century, the shoreline was eroding and the protective high dunes began to disappear, probably related to sea level rise. A couple of passing storms flooded gardens and a house or two, and people began to worry. Then came the December 1899 storm, perhaps the storm of the century in this region. The old folks said it was the worst storm ever (still a common observation of community elders in just about any hurricane that makes a direct hit). Fishing boats were destroyed. Most of the houses were flooded. All the gardens were destroyed, and even the cemeteries were devastated. Apparently no one was killed, but life in Diamond City could never be the same.

People began to move away, most to Harkers Island, which had formed behind Diamond City during a higher sea level period about 125,000 years before. Other buildings were moved to nearby Morehead City on the mainland. Some houses were completely disassembled and moved off as piles of lumber. Others were moved whole or in halves perched on fishing boats powered by sails. It was a process requiring the participation of most of the males in the community; they received payment in the form of a bountiful communal meal.

Three years after the Great Hurricane of 1899, Diamond City was no more. Today most of the house sites are well out at sea on the continental shelf, the high dunes used by whale spotters are completely gone, and the shoreline is retreating at 10 feet (3.0 m) or more per year. Many communities have been impacted by this two-step process in which rising sea levels (step 1) allow storms to remove the protective dunes, ultimately allowing the storm impacts to penetrate farther onto the island and threaten the existence of the town (step 2).

Thompson's Beach, New Jersey (1950)

Thompson's Beach (and nearby Moore's Beach, which has an almost identical history) is an island along the New Jersey shore of Delaware

Bay. The waves along such shorelines are much reduced because there are only short distances over which winds can generate waves. Thus, when the big storms occur, they often are more memorable for their floods, more properly called storm surges, rather than for the size of their waves. At its peak, the village on Thompson's Beach consisted of eighty-eight buildings, mostly second-home seasonal dwellings. But the joys of beach living ended abruptly with arrival of the Great Flood of November 25, 1950. Property damage from wind and rising water was extensive; eight people died and six more went missing. The village was decimated, and only three or four houses remained on their foundations.

By 1998, the last two year-round residents were forced to leave the island and the remaining buildings were torn down. The local government had refused to further maintain the mile-long road to the island that coursed through the marsh. It had become increasingly costly to maintain, due in part to increasingly frequent flooding, a reflection of a rising sea level. Today, a walk to the island must be scheduled for low tide.

Baytown's Brownwood Subdivision (1983)

In the 1950s, an upscale Baytown, Texas, community known as the Brownwood subdivision was constructed on land jutting into the San Jacinto River at the head of Galveston Bay. Roads, streetlights, and sewer and water lines were put in place to serve four hundred homesites on the 450-acre peninsula that lay almost in the shadow of the massive Baytown petroleum refinery.

The combined use of groundwater by the refinery and the town of Baytown caused the land to subside. Similar problems are today so widespread south of the city of Houston that the region is subdivided into subsidence districts designed to control the rate of groundwater extraction. While other parts of the nation are divided into agricultural districts, water districts, or mining districts, Baytown is part of a subsidence district.

By the early 1970s, more than three hundred houses had been built at the Brownwood site even though it had already become obvious that the land was subsiding (or the sea level was rising) at a very rapid rate. Today, sea level at Brownwood is about 8 feet (2.4 m) higher than it was in 1951!

Again, storms were the agents that precipitated action to respond to sea level rise. The gravity of the situation was brought home to Brownwood residents as early as 1961, when a storm flooded the community and required its evacuation. With the floods in the wake of Hurricane Alicia twenty-two years later, it was abundantly clear that the community had to be abandoned. Eventually, the Federal Emergency Management Agency (FEMA) moved most of the houses to higher ground, and for a decade or so the Brownwood subdivision remained a strange ghost town, with streetlights and telephone poles lining roads and driveways to nowhere.

Then, in a 1995 move that may presage the future of many current coastal settlements, the Brownwood subdivision was turned into a wetland and wildlife refuge. Much like Thompson's Beach, New Jersey, Brownwood is now part of a large salt marsh restoration project. Concrete streets and driveways were broken up and used to build a seawall along the tip of the peninsula, and trees were planted on the remaining high ground. All told, about a hundred acres of new marsh were created from the land that once was the site of Brownwood's post–World War II suburban yards and houses.

Groundwater extraction is responsible for large sea level rises in many other parts of the world as well. Perhaps the most spectacular example is the Koto district in eastern Tokyo where industrial overuse of water has caused a 31-square-mile (80 km²) area to sink as much as 8.2 feet (2.5 m) below mean high tide level. Two million people live in this subsidence zone, protected by large levees that surround the city. Another very prominent example of subsidence is found in the cities of Venice and Ravena in Italy's Venetian Lagoon along the Adriatic Sea. Both suffer frequent flooding due to an already elevated sea level, again caused mainly by groundwater removal. There is also

One of the few houses that remained from the three-hundred-house Brownwood sub-division of Baytown, Texas, along Galveston Bay. The sea level rose 8 feet above its 1950 level because the oil refinery (part of which can be seen through the window) extracted water from collapsible geologic formations. The streets, sidewalks, and all buildings have today been removed, and the Brownwood subdivision is now a salt marsh public park. (Photo by Joe Kelley)

local subsidence in New Orleans, which lowered the elevation of some of the levees. While sea level rise from subsidence is not a consequence of sea level increases due to global warming, the effects are essentially the same.

Sea Level Rise and Nature Preserves

Rising seas will affect not just the built environment, but the natural world as well. Some of the natural environments that will be impacted, such as the Sundarbans, the Great Rann of Kutch, and the Everglades, are unique environments, and the animals unique to these ecosystems will have no place to migrate and thus may perish.

The Bering Land Bridge National Preserve consists of 2.7 million acres and sits astride the Arctic Circle. The Shishmaref Island Chain, part of this preserve, sits along the Chukchi Sea at the tip of Alaska's Seward Peninsula. (We discussed the trials and tribulations of the village of Shishmaref in chapter 1.) Sea level rise in this and other Arctic preserves operates in tandem with greatly increased storm impacts because of longer periods of ice-free conditions on the ocean and melting of beach permafrost.

The Sundarbans of both India and Bangladesh is part of the Ganges Delta. It contains one of the world's largest mangrove forests, discussed in Chapter 6, and is the home of the endangered Bengal tiger. It is also home to a number of other endangered species, including marine turtles, crocodiles, and freshwater dolphins. Two million people live on the food (fish, crabs, mollusks, and honey) and firewood from the Sundarbans. The very fact that this is a mangrove forest is an indication of its close proximity to sea level and the great peril it faces from sea level rise. According to World Bank estimates, a 24-inch (60 cm) sea level rise would inundate the whole area and a 3.3-foot (1 m) rise would destroy the Sundarbans.

The Great Rann of Kutch in Gujarat, India, is the largest wildlife reserve in India and a World Natural Heritage Site. It is home to the last two thousand endangered Indian wild asses and to one of the world's largest colonies of greater and lesser flamingoes. The entire habitat, much of it a seasonal salt marsh flat during the monsoon rains, with scattered small islands where the wild asses survive, will likely be submerged by midcentury.

The Cape Hatteras National Seashore in North Carolina is a chain of thin, low barrier islands with a low sand supply. Sea level rise is already narrowing the width of the islands (by shoreline erosion on both sides). Thanks to a lifetime of studies by Stan Riggs, professor of geology at East Carolina University, these may be the best understood barrier islands in the world. Riggs believes the islands may "collapse" or disappear, possibly within the next few decades. With each increment of sea level rise, the possibility of collapse increases. He anticipates that in a storm of sufficient duration and intensity, a large number of new inlets could open up, simultaneously isolating the eight small tourist villages within the National Seashore. If

Riggs is right, a large portion of the Outer Banks will become a long, submerged sandbar quite similar to Louisiana's Isles Dernieres after Hurricane Katrina.

Sadly, the North Carolina Department of Transportation has not heeded this warning. A multimillion-dollar bridge spanning Oregon Inlet, a part of the National Seashore, will be rebuilt in the exact location of the current Bonner Bridge even though it is likely to go to sea within the life span of the bridge.

The Everglades cover more than 1.3 million acres in south Florida. As sea level rises, salt water will intrude on the Everglades and its fauna and flora will change in profound ways. The changes in birds, fish, and various plants are likely to be adverse ones, Everglades National Park Superintendent Dan Kimball has noted. Absent human influence on the planet, the changes would simply be the natural fluctuations that have occurred over millions of years as the level of the seas changed. But with humans involved and the demands of agriculture, tourism, and recreation, the changes are indeed likely to be adverse because there is nowhere for the Everglades to go. The multibillion-dollar Everglades restoration that is under way is designed for a 1-foot (0.3 m) sea level rise in the next century. Since a much larger sea level rise, certainly in excess of 3 feet (0.9 m) in that period, is a strong possibility, it is questionable what "restoration" will mean in the Everglades in the long term.

The Ria Formosa Nature Reserve is the Portuguese equivalent of the Cape Hatteras National Seashore. The reserve is a chain of seven barrier islands in the Algarve (in southern Portugal) near the Spanish border. The area has some unique problems that will make a response to sea level rise difficult. More than two thousand houses on the islands are built on public property and occupied by relatively wealthy squatters, who have even insisted on the right to build seawalls to protect their property. If the government is unable to remove the houses in the future, the place will become a battleground between engineers and sea level rise, to the great detriment of the reserve.

Preparations for moving the Cape Hatteras Lighthouse in North Carolina. In response to the threat of further shoreline erosion and more storms, the lighthouse was moved back 2,000 feet (600 m) in 1999 along the path shown. A twenty-year societal debate preceded the move, which was strongly opposed by local officials, who feared negative publicity about local erosion problems. (Photo courtesy of the U.S. National Park Service)

Although the built-up environments of the coastal areas of the world may capture the lion's share of attention, the natural areas along the ocean shorelines that have been wisely preserved for future generations are clearly also in trouble.

Sea Level Rise and the Peril of Nations

Each coastal nation has unique problems to solve in a time of rising sea level, depending not just on the physical and biological nature of its coasts but also on its resources, politics, and culture. The impact on biodiversity, fisheries, and tourism along heavily developed coasts will be vastly different than in remote, lightly developed regions.

In broad perspective, the principal nation-scale impacts of sea level rise are likely to be the following:

- Loss of agricultural and nonagricultural land
- Flooding
- Increased vulnerability to storm surges
- Accelerated erosion of shorelines and artificial beaches
- Increased salinization of surface and groundwaters
- Increased flood heights of tidal rivers
- Loss of biodiversity (loss of marshes/mangroves/coral reefs)
- Loss of aquaculture, fishery, marina infrastructure
- Tourism decline as beaches erode and resorts are threatened

Land loss itself is generally viewed as the least of the problems brought on by rising sea levels. Of course, the major exception to this view is Tuvalu and other atoll nations that have nowhere to go. More important are the more indirect events that will adversely affect a nation's economic and social well-being, such as patterns of shoreline retreat relative to threatened buildings; loss of tourist infrastructure; loss of coastal infrastructure including seawalls, marinas, utilities, and roads; the salinization of wells; the destruction of sewage and drainage systems with consequent health problems; and loss of the nearshore ecosystem, including edible resources.

The countries with the biggest problems are the atoll nations; deltaic countries such as Egypt, the Netherlands, and Bangladesh; and countries with large, low-lying, heavily developed coastal plains such as the United States, Brazil, and China. The elevation of deltas is always low to begin with, and natural compaction of river muds leading to subsidence adds to the rate of sea level rise. In addition, delta subsidence is increased by extraction of water and oil and by the absence of replenishing sediment trapped by upstream dams. Furthermore, the continental shelves off most deltas are flat and gently sloping, a geometry that leads to storm surges that are particularly high and laterally extensive.

Vietnam

Among developing nations, according to a 2007 World Bank report, Vietnam will suffer enormously in terms of land area lost and people displaced from a 3.3-foot (1 m) sea level rise. The country has two major deltas, the Red River and the Mekong. These densely populated areas are, like the Nile Delta, the breadbaskets of the country. A 3.3-foot (1 m) rise will flood 1,930.5 square miles (5,000 km²) of the Red River Delta and 7,722 square miles (20,000 km²) of the Mekong Delta. Such a rise is expected to displace more than a tenth of the nation's people, gobble up 12 percent of its land area, and reduce food output by 12 percent.

Adding to the problem will be loss of sediment traveling to the deltas from likely future dam construction. China in particular is eyeing the Mekong River as ripe for damming. Loss of sediment will increase the subsidence rate because land will no longer build up; thus, the sea-level-rise rate will increase (just as is happening in the Mississippi Delta today) and shoreline erosion rates will grow in proportion.

Myanmar

The Irawaddy River in Myanmar splits into innumerable distributaries that end up discharging into the Andaman Sea on a 200-mile-wide (320 km) delta front. The Irawaddy delta's rice production has long provided the breadbasket of Myanmar. In May 2008, the catastrophic Cyclone Nargis carved a corridor of destruction across the entire breadth of the lower Irawaddy delta. The cyclone could not have followed a more disastrous route, flooding the deltaic lowlands, which are often less than 3 feet (0.9 m) above sea level, with a 12-foot (3.7 m) storm surge. Perhaps one hundred thousand people died in the storm.

Myanmar's ruling military junta drew worldwide condemnation for its inept response to the human tragedy that was Cyclone Nargis. In Myanmar, the local government was totally unprepared for the storm, even issuing an initial estimate of 390 dead (off by three orders of magnitude). Few warnings were given before the storm arrived, no evacuations were attempted, and no plans were in place for shelters. The widely unpopular and insecure Myanmar military regime couldn't bring itself to quickly loosen its clamp on foreigners, even to allow aid workers immediate access to the huge numbers of homeless and starving people. The saga of this storm response is a vision into the future of vulnerable deltas in incompetently governed countries at a time of accelerating sea level rise. Cyclones like Nargis and other mega-storms will have increasingly disastrous impacts on delta residents as sea level continues to rise.

Bangladesh

The Gift of the Rivers, as Bangladesh is appropriately known, is about 90 percent floodplain or delta and is the poster child of sea-level-rise impact. Around 15 to 17 percent of the country's produce is located at elevations within 3.3 feet (1 m) of sea level. Estimates of numbers of people that will be heavily affected by a 3.3-foot (1 m) sea level rise range from 13 million to 30 million, and at least 15 million likely will be forced to become environmental refugees.

Asia's two biggest rivers, the Ganges and the Brahmaputra, meet in Bangladesh and form the Meghna River, which finally brings the water across the world's largest delta to the sea through a complex pattern of distributaries. The delta's future problems will come from a combination of sea level rise and expected increase in cyclone intensity, which in turn will cause higher storm surges. The 1970 and 1991 storm surges caused by cyclones killed perhaps 500,000 and 140,000 people, respectively, while floods in 1992 and 1998 inundated more than half the country's land area. As sea level rises, the elevation change at the coast will reduce the gradient of the lower river and will thus cause the delta to drain more slowly and floods to penetrate farther inland.

Changes in agriculture and water quality and quantity will be the most obvious everyday manifestations of sea level rise in Bangladesh. Rice production will decrease dramatically. Already, on the outer-most habitable portions of the delta, increased saltiness of the soil has changed crop yields, and a significant part of the rice crop has been lost.

The storm preparation of the democratic Bangladeshi government stands in sharp contrast to the Myanmar military regime. Bangladesh has responded to past catastrophes with coastal building codes, construction of twenty-five hundred concrete shelters high atop pilings, installation of warning systems, organization of a 32,000-person rescue group, and well-publicized evacuation plans. These efforts have greatly reduced the vulnerability of this low-lying country.

Another ray of hope in Bangladesh, however faint, is the diversion of river silt into "soup bowl" depressions on the outer Ganges Delta. Like the transfer of sediment on the Mississippi Delta, the addition of new material adds much-needed elevation.

Egypt

The Nile Delta, perhaps the oldest intensively cultivated region on Earth, makes up only 2.5 percent of Egypt's land area, but the World Bank projects that 9 percent of the country's population will be displaced by a 3.3-foot (1 m) sea level rise. The 3.1-mile-wide (5 km) stretch paralleling the delta coast is mostly below 7 feet (2 m) in elevation and is protected from storm floods and storm wave action by a chain of barrier islands and dune fields along the shoreline. But because of a combination of sea level rise and loss of sand trapped behind the Aswan Dam across the Nile, this protective band of sand is rapidly being lost.

Some officials remain unconvinced that the threat to the Nile Delta is real. Mostapha Saleh, head of Environmental Quality in Egypt, for example, believes that flooding predictions have been exaggerated to draw international attention to Egypt's problem. Saleh was quoted in 2008 in the *Middle East Times* as saying, "If sea levels rise by 1 m that would bring the water inland by about 40 miles, so it is not necessarily a large portion of the delta"! Saleh's comments reflect a head-in-the-sand attitude about sea level rise, not uncommon among politicians in many countries.

The Netherlands

The Netherlands has spent more money and intellectual capital preparing for sea level rise than any other country. Its research institutions are constantly studying the economic, social, oceanographic, engineering, and political aspects of sea level rise. There is also widespread recognition among the citizenry that they are experienced in protecting themselves from the sea and are ready and willing to spend much national treasure to continue to do so. Miles and miles of levees, dikes, nourished beaches and seawalls, groins and jetties, and giant tide gates to control the in-and-out flow of water from the land—all attest to the world's most advanced skill in coastal engineering. Engineering is the solution, say the Dutch, and unlike Bangladesh, they have the money to implement solutions.

The Netherlands is a relatively small and relatively wealthy country. The engineering solutions the Dutch have found for protecting their society from sea level rise would be too costly and too environmentally damaging in most other countries. Many Louisiana politicians have looked to the Netherlands as a model for how the Mississippi Delta coast could be protected. This would be an environmental disaster. It would also be unnecessary. The United States is a big country with plenty of room for its sea-level-rise refugees.

Singapore

Most of the business district, airport, and port facilities of this small island-city-state at the southern tip of the Malay Peninsula lie less than 7 feet (2 m) above sea level. Considerable land has been reclaimed from the sea, and it is particularly vulnerable to rising waters. Like the Netherlands, this is a wealthy, efficient country. But unlike the Netherlands, Singapore has little experience with large-scale coastal engineering. Local politicians assume that extensive diking will save the city, for a while.

Indonesia

Among Indonesia's seventeen thousand islands, the most endangered is Java, where more than 110 million people live. Its capital, Jakarta, with a population of 8.5 million, on the north coast of Java, is Indonesia's largest city and among the most vulnerable cities in the world to sea level rise. Indonesian scientists predict the city's airport will be inundated by 2035. In November 2007, the road to the airport was breached by a storm, an event expected to occur with increasing frequency in the future. The same November storm produced storm surges that stepped over the seawalls of Jakarta along the northern shores. By 2050, about 24 percent of the city will be gone, according to local estimates (assuming no coastal engineering). Some believe the capital will have to be moved 112 miles (180 km) to higher elevations in the city of Bandung.

Sea Level Rise and the Cities

Just as the expanding ocean will affect coastal nations differently, each city that is vulnerable to sea level rise has a different physical situation, including tidal amplitudes, storm surge and storm wave potential, and area of the city that is low enough to be affected by rising waters. The nature of the culture and the resources available to the city will determine the nature of the response. And of course each city will have unique vulnerabilities, such as the subway systems of New York, Boston, London, and Washington, DC. For the world's vulnerable coastal cities, the effects of sea level rise will include the following:

- Blockage of city storm drainage, sewage treatment facilities and subways
- Salinization or pollution of domestic water supplies
- Flooding
- Increase in the extent and penetration of storm surge
- Loss of protective barrier islands that rim many coastal plains
- Infrastructure loss—water, electrical power, roads, railroads, port facilities
- Requirement for dikes, levees, seawalls, and relocation of buildings.

According to a recent study led by Professor Robert Nicholls of Middlesex University and his colleagues and sponsored by the Organisation for Economic Co-operation and Development (OECD), because of expected population increases, as many as 150 million people in the world's major cities—more than three times the 40 million currently endangered—may need to rely on engineering structures such as dikes for survival by 2070 if the sea level rise reaches 20 inches (50 cm) by then, a very conservative estimate.

The OECD study ranked what it considered to be the ten most vulnerable cities in the world as measured by susceptibility of property to flooding (but not storm surge impacts), a useful proxy for evaluating the potential for damage (though not loss or disruption of life) from sea level rise: Miami (the most endangered), New York/ Newark, New Orleans, Osaka/Kobe, Tokyo, Amsterdam, Rotterdam, Nagoya, Tampa/St. Petersburg, and Virginia Beach. Half of the top ten cities are American, and Miami tops this list (and every other list of vulnerable cities).

Using the same criteria, the OECD ranked the vulnerability of major American cities: Miami, New York/Newark, New Orleans, Tampa/St. Petersburg, Virgina Beach, Boston, Philadelphia, San Francisco/Oakland, Los Angeles, and Houston. The top seven vulnerable cities are all East and Gulf Coast communities on the rims of coastal plains. Such rankings should be viewed as order-of-magnitude approximations but are nonetheless useful.

When it comes to the economics of response to sea level rise, it is likely that the problems of the cities will trump those of rural or touristic areas. Preservation of Manhattan will certainly be seen as a higher national priority than preservation of Wrightsville Beach, North Carolina; Ocean Shores, Washington; or hundreds of other American beachfront tourist towns. In the United Kingdom, preservation of London will be higher priority than preservation of Margate and North Blackpool beach resorts. Preservation of the city of Dhaka, Bangladesh, will undoubtedly be higher priority than saving agricultural land on the outer Ganges Delta.

The impact of sea level rise on low-elevation coastal cities over the next century will range from catastrophic to minor depending on the resources a city or nation is willing and able to expend. In addition, much depends on the willingness of more developed countries to aid the developing nations. The 2007 U.N. Bali conference on global warming assumed that the wealthier nations would build sea-walls for the developing countries, but this is questionable. At some point, probably a 3.3-foot (1 m) rise in sea level, the lower elevation rims of vulnerable coastal cities such as Dhaka, Bangladesh; Ho Chi Min City, Vietnam; Lagos, Nigeria; Barranquilla, Colombia; Rangoon, Myanmar; and Abidjan, Ivory Coast, will be abandoned. For some of these areas, perhaps a century from now, mass exodus may prove the only recourse. The same sea level rise will be less damaging in Inchon, South Korea; Vancouver, British Columbia; London, England; Tokyo, Japan; and Miami, Florida, because of available resources in these wealthy countries (if the nations choose to hold the line). But sooner or later, if some of the more pessimistic scenarios become reality, mammoth shifts in population may be required for these cities as well.

— Chapter 8 —

Ground Zero:
The Mississippi Delta

It is time to ask whether pouring billions of dollars into
massive coastal fortifications makes sense. It may be time to
consider whether some areas should be abandoned and left to the
inexorable forces of rising sea levels and ever more intense storms.

BRUCE BABBITT, *Dallas Morning News* Op Ed, September 2008

IN SOUTHERN LOUISIANA, a unique assemblage of natural
and human-induced changes has combined to produce a landscape that has been drowning for decades. This has made
coastal Louisiana ground zero for sea level rise in the United States.
Long before the global alarm bells began ringing about the potential
impacts of warming-induced coastal change, the nightmare of severe
land loss has been all too real for the residents of the Mississippi
River delta and nearby areas. Yet, the Louisiana political establish-
ment continues to believe that coastal communities can remain, even
thrive, in this precarious location.

Historical accounts of the settling of the New Orleans region are
filled with disparaging remarks about the swamps and lowlands that
made life so difficult for eighteenth-century residents. But the area's

importance did not pass unnoticed either: it provided access to what would become America's most critical transportation artery, the Mississippi River. Thomas Jefferson certainly recognized this importance when he bought this "impossible site" from the French in 1803 as part of the Louisiana Purchase. Early French, German, and Spanish settlers occupied the high ground of the natural levees formed by sediment carried downstream by the river. In fact, the nickname for New Orleans, "Crescent City," comes from the shape of the natural levee settled along a bend in the Mississippi River.

In the second decade of the 20th century, engineer and inventor A. Baldwin Wood was able to execute an ambitious plan to drain the city, using large pumps of his own design. Wood's pumps and drainage system allowed the city to expand greatly in area. Today, much of New Orleans and some surrounding areas are below sea level—a fact brought home by flooding from Hurricane Katrina in 2005. As development proceeded over the last hundred years, there was nowhere left to go but onto lower ground.

To many of today's population, the impacts of Hurricanes Katrina (2005), Rita (2005), Gustav (2008), and Ike (2008) have been their introduction to southern Louisiana's extraordinary vulnerability to storms. Much has been made of the role that recent coastal land loss played in the extent of these disasters. Yet, the area has been extremely vulnerable for centuries, long before modern developments. A 1722 hurricane that destroyed New Orleans caused some French colonials to suggest that the area was uninhabitable, and a 1779 hurricane caused a total loss of all buildings and provisions, according to the governor at the time. Early settlements on barrier islands along the delta's outer fringe were almost all devastated by hurricanes at one time or another in their history. Clearly, catastrophe has long been a part of the cycle of life along such a vulnerable coast. Although the famous Pat O'Brien's hurricane cocktail was reportedly named for the shape of the glass (like a hurricane lamp), it seems an especially appropriate name for a drink invented in New Orleans.

Despite the periodic devastation, people stayed. They stayed for

practical reasons—farming, fishing, petroleum, and other business—
and for nonpractical reasons—heritage and unique culture. Writer
Anne Rice explained the post-Katrina resilience this way in the *New
York Times* on September 4, 2005:

> They will rebuild as they have after storms of the past; and they will
> stay in New Orleans because it is where they have always lived, where
> their mothers and their fathers lived, where their churches were built
> by their ancestors, where their family graves carry names that go back
> 200 years. They will stay in New Orleans where they can enjoy a
> sweetness of family life that other communities lost long ago.

Certainly, there are many reasons to attempt the preservation of the
southern Louisiana landscape, ecology, culture, and economy from
the ravages of future storms and sea level rise. The real questions are:
Can we do it? At what cost? Who pays?

Sinking Coastal Louisiana, Rising Sea Level

As discussed earlier, there can be a significant difference between
global, eustatic sea level rise and local, relative sea level rise. In areas
where the land surface is sinking (or subsiding), the actual local rate
of sea level rise can be much greater than the rate of rise on a stable
coast. Such is the case for southern Louisiana.

The lands and wetlands of the Louisiana delta region are formed
by a dynamic interaction of processes that includes Mississippi River
waters, the waters of the Gulf of Mexico, sediment accumulation
from various sources on a growing (or shrinking) delta, the stabilizing
effects of vegetation, and the sea level rise. Geologist R. J. Russell, a
professor at Louisiana State University, recognized in the 1930s the
role these forces play in maintaining the health of the delta's wetlands
and their already rapid coastal retreat. By the 1970s, scientists had
begun to quantify the rate of land loss, and many were stunned both
by the extent of loss and by its rate of acceleration.

Over the last fifty years, the rate of land loss has been estimated

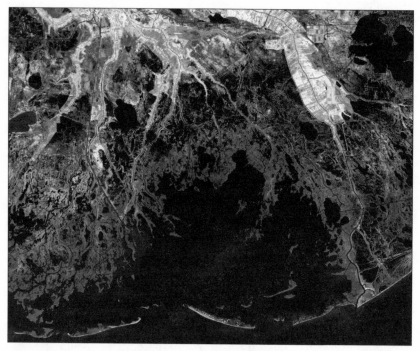

This LANDSAT image of a portion of the Mississippi River delta shows marsh converting to open water. The ragged outline of the marsh is a clear indication of its gradual demise. This is occurring because of sea level rise, land subsidence, and a dearth of sediment coming down the Mississippi River. (USGS)

at a staggering 40 square miles (100 km²) per year. This rate slowed somewhat through the 1990s as measures were taken to reduce the loss and as oil and gas extraction slowed. Still, current estimates place the modern loss at somewhere between 25 and 35 square miles (65 and 90 km²) per year. According to the U.S. Geological Survey, the 2005 hurricane season alone may have been responsible for creating as much as 271 square miles (700 km²) of open water at the expense of wetland. Some of the affected wetlands will slowly recover, but how much is anyone's guess.

In addition, the barrier islands that once ringed the delta are also disappearing. Much of their erosion is the result of storms. The Isles

Louisiana's Chandeleur Islands almost disappeared during Hurricane Katrina. To a large extent, they have not recovered. The arrows point to the same location in these before (2001) and after (2005) Hurricane Katrina photos. (USGS)

Dernieres barrier island chain was severely damaged by Hurricane Andrew in 1992, and the Chandeleur chain was obliterated by Hurricane Katrina. These barrier islands were important habitat in their own right, but they also served to provide wave protection for wetland areas tucked behind the islands.

A tremendous amount has been learned in recent decades about delta processes, wetland ecology, and environmental restoration through the work of researchers on the Louisiana coast who have dedicated themselves to stemming the losses and saving those portions of the ecosystem that can be saved. Any quarrels we may have with specific plans for a restoration of coastal Louisiana do not detract from the tremendous amount of respect we have for the scientists who have helped elucidate the causes and ramifications of the land loss.

River deltas like that of the Mississippi are very large accumulations of sediment transported by rivers to the sea. The Mississippi River delta plain in southern Louisiana encompasses more than a half million acres. The tremendous weight of this sediment causes compaction of the sediments below and also presses down into Earth's deeper layers. The result is an area where land and wetlands are naturally subsiding. This sinking land can be compensated for by the addition of newer sediments on the surface as the river continues to cover the delta plain with fresh sediment. Wetland plants also contribute to the sediment accumulation by dropping leaves and stems (organic sediment). This balance can be upset by a number of factors.

The rapid subsidence of the Mississippi River delta and associated loss of wetlands and barrier islands in Louisiana is a result of a number of natural and human-induced causes that have often reinforced one another. The biggest problem in this land loss, besides the fact that global, eustatic sea level is rising, is a reduction in the amount of sediment being delivered to the delta plain. Before European settlement in the area, the Mississippi River would wag back and forth across southern Louisiana, depositing sediment at its mouth and upstream during floods. This sediment would maintain the elevation of the wetlands and feed the offshore barrier islands. Natural land loss might occur as the river's center of deposition moved from one area to another, but this loss was cyclical and small compared with today's rate of land loss. Over the last hundred years, human activities have starved significant portions of the delta region of this natural sediment replenishment.

Efforts to trap the river within its existing banks and reduce flooding by building levees deprived many wetlands of needed sediment. Of course, such levee building was necessary to protect the growing number of settlers and important infrastructure, presenting a classic Catch-22 situation. In addition, dam building along the Mississippi's tributaries (particularly on the Missouri River) also deprived the downstream Mississippi of sediment now trapped in reservoirs. It is estimated that today the sediment load delivered by the river is half what it was before 1950 when most dams were built. In addition, much of this sediment is funneled out into deeper water by a leveed river channel, rather than spreading out over the wetlands during floods.

This relative lack of sediment deposition profoundly disturbs the balance between subsidence and accumulation on the delta. As a result, many wetlands slowly lower to a level at which the vegetation can no longer survive. In some cases, the marshes are disappearing many miles from the coast.

This situation is made worse by extensive canal building that has taken place for navigation that serves the petroleum industry and other shipping and fishing interests. Canal building results in the direct destruction of wetlands and alters the hydrology to allow the penetration of salt water and storm surges that can also destroy and alter freshwater wetlands.

Some have argued that extraction of oil and groundwater from below the delta plain by various industries has also played a role in increased subsidence. Bob Morton, of the U.S. Geological Survey, has documented the negative impact of extraction-driven subsidence within the Louisiana wetlands. His research causes some consternation among Louisiana politicians because it is more convenient to blame dams on the Missouri (an extrinsic cause of harm) than it is to blame the activities of an important local industry (an intrinsic cause of harm). Recently, some geologists have suggested that some subsidence is a result of tectonic movement of geological units that are buried far below the modern sediments. This may be the scariest prospect of all, because there would be no fix for that problem. The relative contribution of these various factors to subsidence and land loss is still hotly debated.

Finally, the big unknown lurking in the background is the future rate of global eustatic sea level rise and the potential for an increase in storm frequency or intensity. Tide gauge data for Louisiana indicate that locally, relative sea level has risen more than 3 feet (0.9 m) over the last century. Only a small portion of this rise is from a global change in ocean volume. The rest is local subsidence. What if global sea level were to rise 6 feet (1.8 m) over the next century as some have suggested? The local rise in Louisiana, taking subsidence into account as well, could be 9 to 12 feet (3 to 4 m)! This would certainly be catastrophic and nullify most restoration efforts, and directly threaten New Orleans itself.

What Can Be Done about Land Loss?

Louisiana's wetland and barrier island loss is threatening the vitality and integrity of a variety of important biodiverse coastal ecosystems. Industries that depend on a robust coast are also threatened. A diminishing estuary and accompanying wetlands reduces the habitat for commercially important fisheries, and land loss directly threatens the infrastructure for the offshore oil servicing industry.

Proposals to restore coastal Louisiana have been on the table for many years. In recent decades, a plan called *Coast 2050: Towards a Sustainable Coastal Louisiana,* published in 1998 by the Louisiana Department of Natural Resources with numerous partners, provided the first unifying proposal for coastal restoration. This plan was produced under a number of federal and state legislative mandates and attempts to chart a comprehensive approach for reducing the rate of wetland loss and maintaining delta communities. *Coast 2050* has been augmented by the Louisiana Coastal Protection and Restoration Authority's *Integrated Ecosystem Restoration and Hurricane Protection: Louisiana's Comprehensive Master Plan for a Sustainable Coast* (2007) and by the U.S. Army Corps of Engineers' *Louisiana Coastal Protection and Restoration Report* (2008). All of these documents describe a plan that has been evolving over the last few years; its main points can be summarized as follows:

1. Creation of large- and small-scale river diversions. Manage openings in the levees to allow the Mississippi River to spill onto the delta plain occasionally to simulate flooding and to deposit sediment. Such a diversion already exists at Caernarvon where the freshwater wetlands are responding well to the additional sediment.

2. Restoration of marshes using dredged material. Return some sediment that has shoaled in the river or been lost offshore to areas of need by dredging and long-distance piping or other means.

3. Management of navigation channels. Close unused channels to prevent saltwater intrusion, and use some existing channels to move water and sediment to more remote areas of the coast.

4. Restoration of barrier islands. Rebuild the diminishing barrier islands by pumping massive amounts of sand to their shores and stabilizing them with rock structures.

5. Shoreline stabilization. Use seawalls and levees to reduce wave attack along a variety of shoreline types including wetlands.

6. Closure of the Mississippi River Gulf Outlet (the MRGO Channel). This deepwater channel, built in 1965, was believed to have been a primary contributor to the New Orleans flooding from Hurricane Katrina by funneling storm surge waters inland. It is underutilized, and the Corps has recently proposed to dam the channel.

The proposal with the biggest price tag and the most hubris is the "Morganza to the Gulf" hurricane protection project. This is a $10.7 billion proposal by the Army Corps of Engineers and the State of Louisiana to construct approximately 72 miles (120 km) of earthen levees with floodgates designed to protect portions of coastal Louisiana from storms. This is a proposal from an agency that has failed to provide adequate flood protection almost everywhere it has tried. In addition, this project directly contravenes the efforts to protect and restore wetlands. It will be difficult to predict the impact the structure will have on the functioning of existing wetlands, and some have suggested that it will lend a false sense of security encouraging the destruction and development of existing wetlands. It is a project that appears doomed to failure, but that may not stop it.

Unquestionably, there is value in attempting the restoration of certain wetland areas described in the various proposals. Yet, the ultimate goal of preserving, for the long term, the integrity of coastal

The entrance from the Mississippi River in the background to the Caernarvon Freshwater Diversion Project canal. This canal is part of an effort to restore Louisiana wetlands by allowing the Mississippi River water to flow out onto the wetland surface resulting in deposition of sediment. The goal is to build up the elevation of the Mississippi Delta wetlands to counteract sea level rise.

Louisiana at the scale proposed may not be feasible under conditions of continued subsidence and sea level rise. No one pretends that the land loss trends can be reversed. There is even little belief that land loss could be halted completely.

It is important when looking at proposals such as those for New Orleans and the Louisiana delta to distinguish between restoration and engineering. Providing avenues for the river to reoccupy the floodplain so that there is a regular return of sediment to wetlands is restoration. Wrapping barrier islands in rock walls or pumping them full of sand every three years is not restoration. It is engineering. Restoring an ecosystem suggests that at some point constant, active management is not necessary. Engineering structures, by contrast, are rightly assumed to require constant maintenance as well as expansion or alteration. It seems logical that the focus of investment dollars for restoring coastal Louisiana should go to restoration projects that have the best chance to succeed with a minimal amount of engineering. How long can a rock-wall-protected wetland last when sea level is rising and the delta is subsiding?

Two recent reviews have provided thoughtful critiques of Louisiana restoration plans. In 2006, a National Academy of Sciences (NAS) review panel published a report on the *Coast 2050* plan. In 2008, another NAS panel reviewed the Corps' *Louisiana Coastal Protection and Restoration (LACPR) Technical Report*. The *LACPR* draft technical report points out that all plans rely on maintaining the status quo or sustaining the existing landscape. Yet in the judgment of the NAS panel, the Corps of Engineers' report provides no evidence that it will be possible to maintain the current landscape given present and prospective future rates of subsidence, degradation, and, especially, sea level rise.

At the most basic level, there is no analysis of the amount of sediment that will be available relative to the amount that will be required to construct levees and sustain wetlands. In a presentation before the Geological Society of America in 2008, Mike Blum, formerly of Louisiana State University, provided an initial analysis of the sediment available for maintaining Louisiana's coastal wetlands. His results suggest that the sediment simply will not be available to meet the most optimistic coastal restoration goals even without a dramatic increase in the rate of sea level rise. If restoration of the entire coast is not possible, decision makers and citizens ultimately will have to make hard choices about where restoration can be effective and where it cannot.

The NAS report offers the hope that some wetland areas can be maintained into the foreseeable future, but ultimately it is still impossible to tell where the successes will be and where land loss will prevail. On the whole, it is difficult to see how the goal of maintaining a "sustainable" coast can ever be fully achieved.

Former Secretary of the Interior Bruce Babbitt put it this way in 2008 (as reported by *Science News*): "The ineluctable fact" is that within the life span of some people alive today, "the vast majority of that land will be under water." He also faulted federal officials at the time for not developing migration plans for area residents and for not having the "honesty and compassion" to tell Louisiana residents the "truth."

Will Restoration Provide Storm Protection?

The damage inflicted on the Mississippi Delta by Hurricanes Katrina, Rita, and Ike has led many to argue that the disappearing wetlands and barrier islands are greatly increasing the exposure of delta communities to storm damage. There is no question that the degradation of Louisiana's spectacular freshwater wetlands is an environmental tragedy of grand proportions. The loss of habitat, species richness, and inaccessible beauty is disheartening. Many proposed wetland restoration efforts can be justified by the simple desire to preserve this critical ecosystem.

But economic arguments sell better. And certainly, continued wetland loss threatens valuable fisheries, hunting and trapping of valuable fur-bearing animals, and the livelihoods of others who directly utilize these coastal ecosystems. Yet, these benefits would seem even grander if it could be shown that wetlands provide significant storm protection for communities, infrastructure, and even the city of New Orleans. Then the benefit/cost ratio of coastal restoration projects would improve dramatically.

For this reason, there is growing focus among state officials, some scientists, and even environmentalists on the idea that wetlands and barrier island restoration can save coastal Louisiana and ultimately New Orleans from storm damage and coastal flooding. This is a false promise in our view. Nevertheless, it is an attractive sales pitch for a multibillion-dollar restoration project. If one can argue that oil industry infrastructure can be protected along with the city of New Orleans, the benefits of restoration seem limitless. The suggestion that culturally important communities may be able to remain intact in the delta region in a time of rising sea level is also a strong selling point to those residents who would never want to live anywhere else.

Yet, there is little evidence that restoring wetlands will protect infrastructure and many delta communities. Hurricanes repeatedly flooded and inundated the area in the eighteenth century, before major land loss began. The structures in those days were certainly less resistant to storms than they are today, so comparison is not direct, but it is obvious from historical accounts that the impacts of coastal storms reached well inland even during a time when Louisiana had a "healthy" coast.

Storm surge is the elevated water level associated with the passage of a coastal storm. It is like a very high tide. The height of the storm surge at any point along the coast is dependent on a number of factors, such as the shape of the coast, the offshore slope, the track of the storm, and the storm meteorology (wind speed, size, etc.). During Hurricane Katrina, storm surges along the Mississippi coast reached 30 feet (9.1 m) in height. Imagine having a high tide that is 30 feet (9.1 m) higher than normal. Now, imagine there are hurricane-driven waves on top of that extra-high tide. That is the impact of storm surge.

Recently, it has been suggested that the primary reason for restoring wetlands in coastal Louisiana is to reduce the impact of storm surges that threaten coastal communities. It makes intuitive sense that more wetland in front of a community should equal lower storm surge, but the relationship between wetlands and storm surge is not that simple. The highest storm surge measured after Hurricane Hugo (1989) in South Carolina was in the town of McClellanville, which is "protected" by a barrier island and almost 6 miles of marsh. The surge was actually higher behind the marsh than in front of it.

Of course, there are also some areas where marsh seems to have provided some storm surge reduction, but not a significant amount. Ty Walmsley of the Corps has shown that the same wetland in Louisiana had somewhat reduced storm surge during one storm, but storm surge increased across the wetland during another storm. Even the storm surge reduction he did find was fairly small in the grand scheme of things (around 1 foot). This is not significant storm protection for a low-lying coast.

There is no solid scientific field evidence elucidating the relationship between wetlands and storm surge or wave attenuation. The widely quoted Army Corps of Engineers' assertion that you get a 1-foot (0.3 m) reduction in the height of a storm surge for every 2.8 miles (4.5 m) of wetland has no foundation in science. The ratio comes from a 1963 report using data of dubious quality stretching back to 1915. The data represent only a handful of points from seven different storms. One can't tell if the areas measured were truly wetlands or included some dry land. The line that was statistically fit to the data in order to come up with the ratio of 1 foot to 2.8 miles (0.3 m to 4.5 km) is a poor fit to the data, and many Louisiana scientists now repudiate both the data and the use of the numbers. Yet, a simple Google search indicates that these data are still widely reported by the media, government agencies, and environmental groups as a quantifiable benefit of wetlands.

More recently, sophisticated computer models have been used to describe the potential impact of wetlands on reducing storm surge. Some have suggested that these models provide rudimentary evidence that wetlands can have an impact on lowering storm surge heights. The problem with these models, or with any study of storm surge and wetlands, is that there are almost no data to use to calibrate the models, to check their predictions, or even to do simple quantitative analysis.

Storm surge data is particularly difficult to collect. It cannot be done via satellite, and you can't go personally to measure it during a storm for obvious reasons. That leaves field surveys of storm surge markers, such as a still-water line. The problems with these proxies are that they are only estimates of the storm surge height and there are typically only a limited number of them to measure. Missing are measurements over open wetland areas because there are no places for

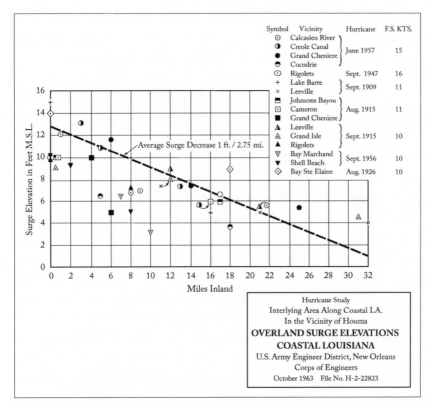

Symbol	Vicinity	Hurricane	F.S. KTS.
☉	Calcasieu River		
◖	Creole Canal	June 1957	15
●	Grand Cheniere		
◗	Cocodrie		
☉	Rigolets	Sept. 1947	16
+	Lake Barre	Sept. 1909	11
×	Leeville		
▣	Johnsons Bayou		
⊡	Cameron	Aug. 1915	11
■	Grand Cheniere		
▲	Leeville		
△	Grand Isle	Sept. 1915	10
▲	Rigolets		
▽	Bay Marchand	Sept. 1956	10
▼	Shell Beach		
◇	Bay Ste Elaine	Aug. 1926	10

Average Surge Decrease 1 ft. / 2.75 mi.

Surge Elevation in Feet M.S.L.

Miles Inland

Hurricane Study
Interlying Area Along Coastal LA.
In the Vicinity of Houma
OVERLAND SURGE ELEVATIONS
COASTAL LOUISIANA
U.S. Army Engineer District, New Orleans
Corps of Engineers
October 1963   File No. H-2-22823

This graph, taken from an obscure 1963 Corps of Engineers report, is the basis for the assertion that there is a direct relationship between wetlands and storm surge reduction. Viewing this original data, one can only conclude that the relationship is tenuous at best.

debris to get trapped or for a line to be left. How can we examine the impact of wetlands on storm surge when we have almost no quality data of storm surge heights in wetland areas? We can't.

Possibly the best attempt at storm surge data collection to date was begun by the USGS during Hurricane Rita. Scientists rushed to the coast before the storm and strapped forty-seven pressure transducers (for measuring water height) to telephone poles and pilings. These devices were placed at thirty-three different locations and recorded excellent data. Still, these devices were spread over an area of more than 4,000 square miles (10,400 km²), giving little perspective on the variability between the data points. It was a fantastic first effort, though.

The lack of high-quality storm surge data was highlighted in a White Paper released in 2007 at the first Cullowhee Coastal Conference. At this meeting, a panel of around thirty coastal scientists and engineers agreed that lack of storm surge data is the critical issue facing the prediction of coastal hazards. An unspoken ramification of that report is that we have no way of accurately modeling or predicting the impact that the loss or gain of wetlands may have on storm surge.

Which brings us back to Louisiana. Any assertions about the benefits of wetlands for storm surge reduction are quite hypothetical. Some folks have pointed to wetland protection of coastal areas during the Indian Ocean tsunami of 2004. Most of those wetlands were mangroves, which certainly would provide significant protection, but they are not the dominant wetland type in coastal Louisiana. Finally, it should be said that both National Academy of Sciences reports mentioned above conclude that the assertion of a clear reduction in storm surge risk when wetlands are present is dubious.

Yet, it is largely with such assertions that a vast and costly restoration effort is now being sold to the residents of Louisiana and the rest of the United States. This is a misleading, dangerous sales pitch and a perilous promise to make. If coastal communities are lulled into a false sense of security by the promise of future wetland restoration and storm protection, they will never make the hard choices that need to be made in order to protect themselves, their loved ones, and their investments.

The *Louisiana Coastal Hazard Mitigation Guidebook,* published in 2008 by Louisiana Sea Grant, tells it like it is: "The 2005 storms remind us that, try as we might, we will never be fully protected from the forces of nature and we are indeed much more vulnerable to certain disasters than we allow ourselves to admit. . . . Problems arise when we ask the greater society to shoulder our burdens time and again for the same unwinnable positions and the same untenable strategies."

Environmental restoration on the delta is needed and justifiable for maintaining fisheries and ecosystems, but it can't be counted on for storm protection. Conditions in southern Louisiana are likely to only get worse no matter how much money is spent. The next Katrina may be far more costly than the last one. The sooner we admit that protection *cannot* be guaranteed, the sooner we can begin to make sensible decisions regarding the fate of vulnerable communities and the future of vulnerable infrastructure in the delta region. Global sea level rise is coming.

Louisiana coastal managers need to begin developing a Plan B for towns like Cocodrie and even Houma in the lower delta. How can we shift the footprint of those communities as portions of them become threatened? This may sound like defeatism to many politicians, but in our view it is the only way to save these towns over the long run.

Many Louisiana natives speak about how critical it is to keep the unique delta communities together—how this is a culture worth preserving. Planning how residents may be relocated after future storms may be the best chance to keep the communities together. Without a plan for where people can go, they may just scatter throughout the United States as Katrina refugees did. Those who want to preserve the delta culture should be willing to accept that they may not be able to preserve it in place. In this admission may come salvation. This is also true for the people of Shishmaref, Tuvalu, and the many other cultures around the world that are seriously threatened by sea level rise.

— Chapter 9 —

Sounding Retreat

Sea level is rising and the American shoreline is retreating. We face
economic and environmental realities that leave us two choices:
(1) plan a strategic retreat now or (2) undertake a vastly expensive
program of armoring the coastline and as required, retreating
through a series of unpredictable disasters.
J. D. HOWARD et al., *National Strategy for Beach Preservation White Paper*, 1985

O F THE BIG four impacts of sea level rise—shoreline erosion,
storm intrusion, flooding, and salinization—the landward
movement of the shoreline is by far the most attention-
grabbing in most Western countries. We can respond to this erosion in
three ways. We can abandon the beachfront and relocate all buildings
and infrastructure away from the retreating shoreline. We can armor
the shoreline with seawalls, groins, and such. Or, we can bring in new
sand to form an artificial beach and push the beach back toward the
sea. All of these solutions have the same major shortcoming: they
are very expensive. The latter two, armoring and beach nourishment,
have additional limitations. They are temporary solutions, suitable
only for small rises in sea level. Furthermore, because their existence
may encourage an increase in density of development, they may make
long-term response all the more difficult and expensive.

Strategic relocation of infrastructure, or as it is referred to in the

United Kingdom, the surrender option (a less than inspiring char-
acterization), involves moving or demolishing roads, buildings, and
other infrastructure or abandoning them to the sea. Depending on
local and national rules, the relocation option can be very costly or
very inexpensive to taxpayers. If individual property owners are asked
to assume ultimate responsibility for the precarious location of their
investments, relocation can cost taxpayers very little.

Relocation would not be costly to the U.K. town of Happisburgh,
Norfolk, where homeowner Jane Archer discovered that for loan purposes, a bank considered her home atop a rapidly eroding seacliff to
be valued at £1. By contrast, the U.S. government has offered to buy
out threatened beachfront houses at full value in tiny Camp Ellis,
Maine, as an alternative to installing additional shoreline engineering
structures. (The homeowners, however, have refused to sell despite
the frequent damage storms inflict on their homes.)

Surfside, Texas, is an example of how far homeowners will go to
avoid assuming responsibility for their actions. The town is famous
among coastal managers for a beach that is spectacularly strewn with
rock, brick, concrete, and other pieces of wreckage from destroyed
engineering structures once intended to protect houses then perched
perilously on the edge of an eroding ocean shoreline. A number of
houses now reside on the beach, seaward of the high tide line. In
Texas, ocean beach is considered state parkland where houses aren't
allowed, and in this case the state has offered to pay the costs of
moving the houses off the beach and back to safer lots. Fourteen
of the homeowners, however, insist on being paid the price of their
underwater lots before they will allow their houses to be moved. They
claim their plight is the fault of the government, which allowed them
to build there to begin with, and they conveniently find themselves
blameless for their arguably absurd decision to build next to a retreating shoreline.

The construction of artificial beaches (variously called beach nourishment, beach replenishment, or dredge and fill) involves pumping
in dredged sand from an underwater source or trucking it to the

beach. Such beaches are found in Spain, the Netherlands, and many other countries in the developed world. In the United States, mostly on East and Gulf Coast barrier islands, more than three hundred beaches have been replaced, often multiple times. These include Miami Beach, Daytona Beach, and Jacksonville Beach, Florida; Hilton Head Island and Myrtle Beach, South Carolina; Ocean City, Maryland; and Ocean City, New Jersey. These beach renovations are costly, on the order of $1 million to $10 million per mile, and temporary. Most U.S. artificial beaches last less than five years, and so the process must be repeated again and again. Carolina Beach and Wrightsville Beach, North Carolina, have each been nourished more than twenty times since 1965. Supplies of suitable sand in adjacent marine waters are, in many areas (North Carolina and west Florida, particularly), limited.

Because sand placement destroys the nearshore ecosystem, the environmental cost of nourishment is also high. The process kills virtually all the fauna and flora on and within the beach, and such life recovers slowly only to be destroyed by the next sand pumping. Another common problem is the use of poor-quality material in the replenishment (leaving the beach muddy, gravelly, shelly), in which case the original ecosystem never returns and the recreational quality of the beach declines. In areas with nearshore reefs (like much of south Florida), beach nourishment can have a devastating impact on corals, sponges, and reef fish. But protection of buildings, not beaches, is the underlying goal of most beach nourishment.

Artificial beaches are a short-term solution with the term getting shorter. As sea level rises, the frequency of artificial beach emplacement will have to increase for the simple reason that higher sea levels will cause faster removal of the sand. This will increase costs per year substantially and require at some time more distant sources of sand to be tapped, further adding to the expense. Sooner or later, replacement will become economically untenable.

Artificial beaches have another downside as well. They look permanent and may buy time for further high-rise development next

A nourished beach of poor quality on Oak Island, North Carolina. This beach was intended to furnish nesting sites for sea turtles, but incomplete testing of the nearby dredge site led to use of inappropriate beach material. Use of poor-quality sand is a global problem for artificial beaches, which are usually put in place in order to protect buildings from shoreline erosion and not to maintain a beach, whether for tourists, residents, or turtles. (Photo by Andy Coburn)

to the shoreline, a disaster for any future response to sea level rise. Retreat from a shoreline lined with small cottages is certainly less costly (and more practical) than retreat from a shoreline lined with 15-story condos. And retreat we must, somewhere down the line. Beachfront high-rise condos and hotels are the single biggest impediment to a flexible response to sea level rise.

Relocation is likely to be the most economically feasible response to sea level change over the long term. However, coastal communities still seldom give it serious consideration. Although the cost of the move-back alternative will be highly variable, depending on many factors, the argument is often made by pro-development groups that retreat is more costly than beach nourishment. A 2007 newsletter of the American Shore and Beach Preservation Association (ASBPA)

Dune scarp on West Dauphin Island, Alabama. This "dune" of sand containing much mud and pebbles was pumped up from the lagoon behind the island after Hurricane Katrina in 2005 but lasted only a few months--an example of poor engineering carried out by individuals who did not understand the processes of the shoreline. The child playing at the foot of the scarp is in real danger from collapse of the artificial dune. (Photo by George Crozier)

quotes a study claiming that in the state of Delaware, retreat would be four times more costly than beach nourishment for the foreseeable future. The article estimates that the cost of nourishing Delaware's beaches for the next fifty years will be $60 million. A close look at that state's spending, however, indicates that $60 million has been spent for beach nourishment in just the years 2001 to 2007!

In the United States, the ASBPA is the most visible group actively opposing preparation for sea level rise. It is a group of coastal engineering contractors, developers, coastal politicians, and representatives of various coastal business interests. The organization argues that beachfront development should be encouraged and shorelines should be held in place. The group lobbies heavily to ensure that federal and state taxpayers fund this losing battle.

Supporters of this kind of coastal development often assert that coastal economies are too important to walk away from, that they provide jobs and boost the local tax base. But how strong and viable is an economy that requires constant taxpayer support simply to maintain the status quo? Another grand myth is that these oceanfront communities somehow benefit all of us. Most citizens are interested in the beach mainly as a place to have fun. If the buildings are moved back or allowed to fall in when their time comes, the beach will remain wide and healthy as it continues to retreat and fun will still be had by all. The unwise infusion of tax money to "save" beaches is really an infusion of money to save buildings.

In reality, many vulnerable beachfront communities represent a false economy supported largely by federal investment. The value of a lot on any barrier island would be significantly reduced if property owners themselves had to pay to replace the roads and other infrastructure destroyed by the sea every few years or pay true market prices for all types of insurance.

An analysis of a proposed Nags Head, North Carolina, beach nourishment project by the U.S. Army Corps of Engineers indicated that, at least for this location, retreat was four times less costly than nourishment, the reverse of the disingenuous ASBPA estimate for Delaware. Nags Head is one of the oldest beach resorts in the southern United States, and dozens of beach cottages there have fallen victim to the sea in the past. The proposed 14-mile (22.5 km) artificial beach was slated to cost $1.6 billion (more than $100 million per mile) over the next fifty years, assuming that a new beach would have to be pumped up every three years, while outright purchase of all the buildings that were expected to be lost to erosion in the same time frame would cost $400 million.

In early 2009, the Corps made it official that relocating coastal property is more cost effective than trying to protect that property by building artificial beaches. The Mississippi Coastal Improvements Program (MsCIP) was initiated following Hurricane Katrina to "reduce the vulnerability of the region." There are portions of this plan that we disagree with, but those disagreements are overshadowed by the fact that the Corps is finally proposing to buy out coastal properties and relocate public infrastructure.

The cost-and-benefit summary in the MsCIP speaks volumes. A proposal to "restore" the undeveloped barrier islands of the Gulf Islands National Seashore would cost an estimated $477 million. The benefits to the mainland shoreline are estimated at only $17.6 million per year in possible storm damage reduction. So, we would spend almost half a billion in federal tax dollars and we would break even after thirty years (if the restoration is a success). From a scientific perspective, the storm surge reduction benefits of the restoration are dubious, and it is unpalatable to use a national park as an engineered storm buffer.

Yet, in the same document we are presented with a far more enlightened proposed project—the High Hazard Area Risk Reduction Program. The centerpiece of this project would be the purchase of approximately two thousand properties in the most vulnerable locations. The costs of this project are much smaller, at an estimated $187–$397 million, while the benefits are significantly larger at $22 to $33 million per year. This plan could pay for itself in less than six years. Even better, the benefits are guaranteed and long lasting. When the property at risk is gone, it is gone forever. In contrast, the barrier island restoration project cannot guarantee protection and does nothing to get anyone out of harm's way.

Where buildings are considered to be more important than beaches, "hard" shoreline stabilization (seawalls, offshore breakwaters, groins, jetties) is widely accepted as the best erosion solution. However, seawalls are a two-edged sword. They can be more effective than beach nourishment in temporarily protecting infrastructure, but they destroy beaches. The beach will continue to erode after a seawall is placed behind it to protect a building. Over time, the beach will inevitably become narrower and narrower until it disappears altogether.

Furthermore, the beach (and shoreface) will steepen in front of sea-walls as storm waves reflect off the wall. Because the shoreface is steeper, incoming waves touch bottom over a shortened surface. Friction on the waves as they roll in is therefore reduced, and the waves striking the beach and wall become higher and more powerful, on average. Thus, the seawalls will ultimately have to be reinforced and enlarged to withstand the increased wave attack. Where we have long-term experience with seawalls (a century or more on New Jersey barrier island shorelines, many centuries on the Lincolnshire coast of the United Kingdom), we can see that continuous maintenance, enlargement of the walls, and loss of beaches is a fact of life. Our experience in the southeastern United States indicates that natural beaches can be expected to disappear in one to thirty years in front of a seawall. Renourished beaches will typically disappear in less than five years.

Six U.S. coastal states currently prohibit or at least strongly discourage seawall construction: Maine, Rhode Island, North Carolina, South Carolina, Texas, and Oregon. This prohibition is based primarily on recognition that the walls would eventually, as mentioned above, result in loss of the beach in front of them, obviously a blow to the tourist economy. And just like artificial beaches, walls generate a sense of security that encourages intensification of development, making the inevitable ultimate retreat, especially with continuing sea level rise, all the more costly.

Bjorn Lomborg, the Danish global warming skeptic, argues confidently that sea level will rise only 1 foot (0.3 m) in this century. Such a rise is no problem in his view because "almost all nations in the world will establish maximal coastal protection, almost every-where." Lomborg claims that "for more than 180 of the world's 192 nations," the protection will approach 100 percent. Protection must mean seawalls.

Lomborg believes the situation is different for poor countries. Using the example of Micronesia, a particularly low-lying country consisting of numerous atolls, he argues that if nothing is done, the

Massive concrete tetrapod seawall along the shoreline of a tourist community south of Tokyo, Japan. Preservation of beaches is not a high priority in Japan. Instead, prevention of land loss drives shoreline management. (Photo by Jess Walker)

country will lose 21 percent of its land area from the 1-foot (0.3 m) sea level rise; with protection, Lomborg believes it will lose only 0.18 percent of its land. He also argues that if the wealthy nations reduce CO_2 emissions, economic growth would be reduced for Micronesia to the point that they could not afford to build seawalls for protection. The net result is that the country will lose more land area if CO_2 is reduced, an astounding conclusion.

Lomborg's assertions are dubious on numerous counts. The prediction of economic disaster arising from CO_2 reductions rests on so many questionable assumptions as to be unfathomable. The assumption of a mere 1-foot (0.3 m) sea level rise in the next century ignores abundant indications that the rise will be much more. The assumption that Micronesia can afford to seawall its entire shoreline is uninformed. Because of the dependence of the future degradation of the Antarctic ice sheet on offshore topography, it is not clear that a reduction in CO_2 production will actually halt sea level rise. And his assumption that all nations will furnish "coastal protection" on all their coasts is impossible.

Widespread seawall construction as some kind of large-scale solution is not just Lomborg's idea, however. One recommendation of the September 2007 Bali conference on global warming was that wealthy countries should build seawalls for developing countries, much as Japan has built a seawall around the capital city of the Maldives.

The Bali recommendation was rife with irony because the current response to shoreline erosion in the developing world is often more progressive than that of developed nations. In many locations, the response is a highly flexible and low-cost one of moving buildings back when they are threatened. As discussed earlier, in villages on the Niger River delta and on the barrier islands of the west coast of Colombia, for example, houses are built for easy and rapid disassembly, transportation, and reassembly at a new location. There is not enough money to do otherwise. Maybe the developing world has something to teach the developed world!

Worldwide, there are many other considerations that come with building seawalls. As sea level rises, the salinization of the local water supply will drive away atoll dwellers as well as those in coastal Bangladesh and the Bahamas, even if they were successfully protected from wave attack by walls.

As was the case with the flooding of New Orleans by Hurricane Katrina, when storms overtop seawalls, as they inevitably must, any community surrounded by walls will become a pond. The storm surge of Hurricane Ike (2008) topped the 17-foot-high (5.2 m) Galveston, Texas, seawall, the mightiest barrier island seawall anywhere. Most of the city's flooding, however, came from the bay side.

In the United States, the entire 3,200-mile (5150 km) barrier island stretch running from the south shore of Long Island to the Mexican border is eroding at rates that typically range from 1 to 5 feet (0.3 to 1.5 m) per year. More than half of this length is in Florida, the state least prepared to respond to sea level rise. The next longest is North Carolina's 300-mile (480 km) barrier island ocean shoreline, about

A house on Soldado Island on the Pacific coast of Colombia, in the process of being moved to another site. At the time this photo was taken, the shoreline was moving back about 3 feet (0.9 m) per month, probably due to a very rapid sea level rise caused mainly by subsidence (sinking) of the island. The houses are constructed using panels, one of which remains on the house in this photo, that are light enough to be carried by hand. (Photo by Bill Neal)

135 miles (220 km)of which is developed. Coastal protection from a rising sea level for barrier islands would require walling entire islands on all sides. In addition, what will we do with the thousands of miles of estuarine, bay, and lagoon shorelines behind the barrier islands? What about low-lying deltas like that of the Mississippi River?

If we choose to hold the line as sea level rises, the total length of seawalls that would have to be constructed and maintained is daunting indeed. The estimated length of American shoreline of the contiguous United States, including bays and the open ocean, comes to 54,500 miles (87, 700 km). Including Alaska and Hawaii brings the total to 89,500 miles (144,000 km).

The cost of building seawalls is highly variable from place to place. Massive walls such as the one lining the open ocean shore at Galveston,

Texas, completed in 1910, will cost in excess of $10,000 per foot to replace. The O'Shaughnessy seawall in Ocean Beach, California, would cost $5,500 per foot to replace. Small walls or bulkheads designed to prevent shoreline retreat, but expected to be overtopped in big storms, would cost on the order of $300 to $1,000 per foot of shoreline.

If we assume $5,000 per foot cost of open ocean shorelines, half that for seawalls on the bay sides of barrier islands, and $500 per foot to stabilize estuarine shorelines, the total cost of initial wall construction in Florida (where most shorelines are lined with high-rises) alone would be on the order of $50 billion for the barrier islands and $30 billion for estuarine shorelines. For North Carolina (where single-family homes and small condos dominate the beachfront), construction of seawalls on the developed portions of barrier islands could be $5 billion, and walling the state's long estuarine shorelines could be $9 billion. These are truly "back of the envelope" estimates but are likely of the right order of magnitude.

There is a further complication as well: on all types of sandy coasts, unwalled shorelines on either side of walled shoreline reaches would continue to retreat. The result of such differential rates of erosion would be that walled communities would, over ensuing decades, become "capes" protruding out to sea and subjected to significantly higher waves.

'Round the World

All around the globe, there is growing awareness of coming sea level rise. To date, the United States appears to be behind in what are still very preliminary efforts of many other countries. In 2008, the EPA released an important document intended to set the stage for the nation's response to sea level rise, but the stated goal of the report was to add to the nation's prosperity while responding to sea level rise. Maintaining prosperity may be desirable, but you can't have your cake and eat it too. A report with such conflicting goals cannot be taken seriously. Response to a major sea level rise will, of course, involve economic sacrifice on the part of property owners, government, and society as a whole even though jobs will be created in building relocation and other industries.

Initial major sea-level-rise impacts on U.S. development will likely occur along our barrier island coasts. Eventually, urban problems, especially stormwater and wastewater disposal, will begin to take precedence over preservation of beach communities. When our main population centers are truly threatened, and we begin to build dikes and move ports and other infrastructure, small beachfront communities are likely to become declining public priorities. The end result, decades from now, but certainly in this century, will be abandonment of many island tourist communities and, unfortunately, massive seawalling of others.

Today in the United States, action on sea level rise occurs in scattered pockets on a mostly local scale. In Olympia, Washington, a controversy erupted over the siting of the new city hall. Detractors argued the planned site was on low-elevation land built out into Puget Sound and was sure to be inundated within a few decades. A new site at higher elevation was chosen. In Santa Barbara, California, a citizens group proposed to paint a blue line around the city at the 23-foot (7 m) elevation contour to show a worst-case scenario of sea level rise (melting of the Greenland ice sheet). The voters threw it out. Joseph Riley, the mayor of Charleston, South Carolina, said that replacement and upgrading of the city's stormwater drainage system was a necessity because of rising sea levels.

Florida takes the prize for being the least prepared of all, especially given its extreme vulnerability. The state effectively has no building setback-from-the-shoreline requirement, and yet it is bound to experience numerous powerful hurricanes in coming decades, storms that will have been intensified by global warming, and it has hundreds of miles of high-rise-lined beaches. When insurance companies backed away from insuring coastal property in 2006, the state, instead of taking advantage of the opportunity to slow down hazardous development, simply decided to take over the financial obligation—in spite of the fact that a couple of hurricanes in succession could wipe out the state's treasury. In early 2009, State Farm Insurance decided to pull out entirely from insuring property in Florida after state regulators denied a request for a rate increase. At a time when sea level rise should be a part of every coastal management program, Florida seawalls are growing by the day. The cost of beach nourishment in the state is skyrocketing, as is the number of people at risk from future storms.

Coastal engineers in the state blame most of Florida's coastal erosion on coastal engineering activities that interrupt sand transport along beaches rather than on sea level rise. In addition, coastal engineering consulting companies make their living from beach nourishment projects. Why would they ever want to advocate for relocation of buildings? Meanwhile, coastal high-rise construction continues, even though, in a purely geometric sense (because it is so low and flat), Florida is the most endangered state in the nation from sea level rise and Miami the most endangered major city.

Lighthouses, the sentinels of the sea, have a long history of falling into the ocean due to shoreline retreat. Located, as they are, as close to the sea as possible, they are global symbols of the destruction ahead for our society as we face a future of rising seas.

The Eddystone light in Plymouth, United Kingdom, famous in song and legend, went through five iterations between 1696 and 1882 when the current version was built. Each successive lighthouse was destroyed by storm or fire. When the Great Storm of 1703 took out the first lighthouse, its overconfident designer, Henry Winstanley, a famous architect of the time, was inside making repairs. He was never seen again.

Among the U.S. victims of the sea and shoreline retreat are Ponce de Leon Lighthouse, Florida, lost in 1835; successive lighthouses on Sandy Point, Block Island, Rhode Island, in the 1830s and 1840s; Chatham, Massachusetts, twin towers, 1879 and 1888; Cape Henlopen Light in Delaware in 1926; and Tucker Beach Lighthouse,

A high-rise-lined shore on Florida's west coast. Two of the buildings shown here are still under construction. Siting of high-rise buildings immediately adjacent to an eroding beach is not only irresponsible, it also prevents any flexible response to sea level rise. (Photo courtesy of NOAA)

Long Beach Island, New Jersey, 1927. Cape May, New Jersey, is now on lighthouse number three, and perhaps most startling of all is the Morris Island, South Carolina, lighthouse, still standing tall and surrounded by the ocean, 1,600 feet (500 m) offshore.

The Morris Island light replaced the original lighthouse that was pulverized by Yankee cannon fire during the 1863 attack on Ft. Wagner in the Civil War (reenacted in the final scenes of the movie *Glory*). Built in 1876 about 1,600 feet (500 m) behind the beach, it now stands well offshore, leaning at an angle of 2 or 3 degrees. Its foundation is believed to be at least 35 feet (11 m) deep, which may be why it has been able to survive in shallow water in the open ocean.

The saga of the Cape Hatteras Lighthouse has all the elements that will face society as a whole in a time of sea level rise. This lighthouse was also built 1,600 feet (500 m) back from the shoreline but the base of its foundation was only 6 feet (2 m) deep, well above sea level. This meant that had the shoreline moved past the lighthouse, the structure would have toppled over. First lit in 1880, this black-and-white spiral-striped lighthouse, the second on the site, had become virtually the symbol of the state of North Carolina by the mid-twentieth century. In a 1980 storm, waves swirled around the base of the structure, and it appeared to be on its way to join the list of felled lighthouses. It was saved by a quick-thinking National Park Service that tore up a nearby parking lot and threw it into the roiling surf while the storm still raged. The erosion-threatened structure was finally moved 2,000 feet (600 m) back from the eroding shoreline in the summer of 1999.

The move came after an onerous two-decades-long societal debate that holds lessons for erosion-threatened communities everywhere. Even though the structure was "protected" by a steel groin and a large sandbag seawall, it was clearly in danger of falling into the sea in a big storm. It was also clear that sea level rise was a factor in the shoreline retreat. The Outer Banks of North Carolina are currently thinning at a rapid rate (several feet per year), as are most of the world's coastal plain barrier islands, by erosion on both sides, a sure sign of an expanding ocean. Beach nourishment at Cape Hatteras had been tried but it didn't take, and the large seawall approach was rejected by the National Park Service, which had recently adopted a policy of letting nature roll on at the shoreline.

Despite its obvious precarious existence, there was strong resistance in local communities to moving the lighthouse. "We will be the laughingstock of the coast if we move the lighthouse," was one often-stated platitude. Experts were found who testified that the lighthouse would fall apart if it were moved and others who testified that the lighthouse was in no particular danger from the sea. Politicians refused to seek federal money for the move. Local proponents of the move were ostracized, and violence was threatened against workers who might be involved in the relocation.

Local resistance may have stemmed primarily from a concern of local politicians that the move would draw international attention to the erosion problem, which might hurt local real estate sales. Some may have been worried that if the idea weren't stopped at the outset, relocation of structures would become the norm for all oceanfront development. We wish this were the case. The eventual move certainly brought the expected international attention, but real estate sales, especially for beachfront property, continued to prosper.

The logjam of resistance to the move was finally broken by a joint National Academy of Engineering and National Academy of Sciences panel, which concluded that the lighthouse had to be moved if it was to be saved and that moving it could be accomplished by off-the-shelf technology. They were right.

Other countries are developing a variety of approaches to respond to the expanding oceans. Over the last few decades, development along the Spanish coast outstripped the rest of Europe's. The result has been a nightmare of pollution and ugliness. In Spain's coastal region, there are 273 towns with four million residents without waste-water treatment. In 2007, the country took 365 beaches off an approved swimming site list rather than make efforts to clean them up. Dozens of new marinas have sprung up, and the coastal region has become a haven for mafia, drug traffickers, and money launderers, among other shady elements.

In 2008, the new socialist government declared that the time had come to enforce the 1988 Ley de Costas (Coastal Law) that made it illegal to build within 328 feet (100 m) of Spain's 3,100-mile (5,000 km) coastline. Anticipated shoreline retreat due to sea level rise is a prime motivation behind this drastic move, although the government's estimate that erosion along the Mediterranean coast by 2050 will extend 60 feet (18 m) inland seems very conservative. Spain's $7 billion plan, which may affect as many as 500,000 people, is to purchase, relocate, and in some cases demolish buildings that were built after 1988 and flouted the law. Politically, this drastic move is less problematic than it might seem because many of the homes threatened by the removal order are owned by nonvoting foreigners, most of them British and German retirees.

South Australia's Yorke Peninsula, which lies between Spencer Gulf on the west and Gulf St. Vincent to the east, is the site of a landmark decision in the saga of human response to sea level rise. At Marion Bay, near the southwestern tip of the peninsula, Northscape Properties planned to site eighty new homes on approximately 20 acres of land near the shoreline. South Australia development requirements are that buildings along the coast must be set back to where they will not be affected by erosion for the next hundred years. In addition, a coastal reserve of 165 feet (50 m) wide must be preserved for recreation and ecological protection. It was clear that the Northscape development did not meet the law's requirements. The Yorke Peninsula Council listened to several experts and concluded that sea level would rise about 1 foot (0.3 m) in the next fifty years, the shoreline would retreat 115 to 130 feet (35 to 40 m) over the same time frame, and this erosion would put the shoreline at the edge of the seawardmost lots. The South Australian Supreme Court backed up the local government's insistence that sea level rise was a basis for restricting development—a first for Australia and perhaps for the world.

The United Kingdom has wisely divided up the English and Welsh coasts into eleven units of manageable length based on geology (called sediment cells) independent of political boundaries. Shoreline management plans are then devised for each cell. It is widely accepted that all of the shoreline can't be protected and that houses and even some villages will have to be surrendered to the sea.

Abandonment has been recommended for 25 square miles (65 km²) of the flat, low-elevation Norfolk "Broads" along with six villages and five small lakes in Norfolk County, East Anglia. Lady Barbara Young, the chief executive of the U.K. Environment Agency, says that between salinization and greatly increased storm damage potential, the sea should be allowed to breach the shoreline, flood the flats, and form a new bay. Coastal experts believe that seawalls would fail to hold back the shoreline here in a time of rising sea level. Most of the villages date from before medieval times, with historic buildings and bridges still standing, making the decision (not final yet) most difficult. There already is a long history of flooding in the area, including the record of a 1287 storm surge flood that killed nearly two hundred people in the village of Hickling. A few miles to the north of the Broads, the village of Overstrand, a cluster of 135 houses plus several businesses, will also be allowed to fall into the sea in this century. Abandonment of this area had been considered in 1662 after a great storm. Local citizens objected strongly, and as a result two thousand men were press-ganged to build new dunes—an approach that obviously can no longer work.

The Thames River Barrier, on the outskirts of London, was designed to protect $160 billion worth of buildings and infrastructure in London from storm surges moving upriver from the sea. The gates have been closed 108 times since they became operational in 1982, but since they were designed in the 1970s, the local sea level rise rate has doubled and it is assumed that it will accelerate even more. With a modest rate of sea level rise, the gates may remain useful until 2082, but with an extreme rate of sea level rise, the gates' capacity may be exceeded as soon as 2020. New design efforts are under way.

The Dutch, masters of protection from the ravages of the sea, have sheltered themselves in recent years with a variety of storm surge floodgates similar in function to the Thames River Barrier. The two grandest ones are the gates protecting Rotterdam and those across the Eastern Schelde, the latter of which was not designed with sea level rise in mind. If breeched by a storm surge, a large developed area of the Eastern Schelde River distributary valley would flood. In any event, a significant sea level rise of 3 feet (0.9 m) or more will probably negate all of the world's current storm surge floodgates.

The Rising Sea Revisited

Here we give a summary of our most salient conclusions and recommendations and our views on decisions we believe need to be made to begin to address the challenge of sea level rise.

The causes of sea level rise are changing. The relative importance of the various drivers of sea volume change (eustatic sea level change) will likely be different in this century than they were in the twentieth century. According to the IPCC, in the twentieth century, the major contributors were, in descending order of importance,
- Thermal expansion of the oceans
- Mountain glacier melting
- Melting of the Greenland ice sheet

Recent data from a variety of sources suggest that in the twenty-first century, the West Antarctic ice sheet will likely become the major source of meltwater, possibly replacing thermal expansion as the most important cause of eustatic rise, making the main drivers these:
- Melting of the West Antarctic ice sheet
- Melting of the Greenland ice sheet
- Thermal expansion of the oceans
- Mountain glacier melting

In the short term, reduction of carbon dioxide emissions will not halt sea level rise. Many researchers have pointed out that global warming and sea level rise have a certain momentum. Even if we slow the input of CO_2 into the atmosphere, there will likely not be a direct reduction in the rate of sea level rise. The potential collapse of additional ice shelves in Antarctica, for example, is an unpredictable wild card that could raise sea level even as CO_2 falls. We must expect that rapidly rising sea level will be with us for a long time. By no means does this imply that reducing anthropogenic CO_2 should not be a critical long-term goal for our society.

It's not just sea level rise. Storm surge, storm waves, shoreline erosion, groundwater salinization, and infrastructure destruction will force a retreat from the shoreline long before actual inundation occurs. Simple maps showing the areas that will be slowly inundated by a given amount of sea level rise should be viewed with great skepticism. They do not truly consider the inland reach of the changes to be wrought by rising sea level.

Assume a minimal sea level rise of 7 feet (2 m) by 2100 for planning purposes. This is not a prediction; it is a scenario, a recommendation. But a rise of this magnitude (7 ft; 2 m) is a real possibility. Seven feet is a catastrophic sea level rise. At a bare minimum, we recommend using a minimal 3 feet (0.9 m) for a 50- to 100-year planning horizon in communities where the politics would not permit the consideration of more forward-looking coastal management.

Three feet of sea level rise will doom much, if not most, barrier island development. Maintaining the static shoreline required to keep sandy barrier islands in place is economically, environmentally, and oceanographically impossible with a 3-foot (0.9 m)rise in sea level.

Immediately prohibit the construction of high-rise buildings in areas vulnerable to future sea level rise. Decisions concerning community planning and development should be based on minimizing or avoiding altogether the damage from the expanding ocean. This means, first and foremost, no more high-rises near the beachfront. Buildings placed in future hazardous zones should be small and movable or disposable.

Relocation of buildings and infrastructure should be a guiding philosophy. Instead of making major repairs on infrastructure such as bridges, water supply, and sewer and drainage systems, when major maintenance is needed, go the extra mile and place them out of reach of the sea. In our view, no new sewer and water lines should be introduced to zones that will be adversely affected by sea level rise in the next fifty years. Relocation of some beach buildings could be implemented after severe storms or with financial incentives.

Stop government assistance for oceanfront rebuilding. The guarantee of recovery is perhaps the biggest obstacle to a sensible response to sea level rise. The goal in the past has always been to restore conditions to what they were before a storm or flood. In the United States, hurricanes have become urban renewal programs. The replacement houses become larger and larger and even more costly to replace again in the future. The problem is compounded when even people whose rental investment houses were destroyed are considered victims; maybe people who insist on building adjacent to eroding shorelines facing the open ocean should be considered fools rather than victims?

Underlying this problem in the United States is the Stafford Act. Passed by the U.S. Congress in 1988, the act allows the expenditure of money to restore community infrastructure once the president issues a disaster declaration. It is an unquestioning, automatic response to a disaster. No environmental impact statement is required, and money may even be spent on coastal areas where other expenditures of federal money are prohibited. Those who invest in vulnerable coastal areas need to assume responsibility for that decision. If you stay, you pay.

Stop asking only coastal engineers for a solution to coastal erosion. If a coastal community asks a coastal engineer for a solution to a coastal erosion problem, that community gets a coastal engineering response (e.g., build a seawall, build a groin, renourish the beach). Coastal engineers are selling a product. They are not likely to suggest that the community relocate property. This would put them out of business. Beach nourishment, the currently preferred method of fighting coastal erosion, is becoming increasingly expensive. In the future, beaches will need more sand, more frequently. In most cases, the sand resources are not available to fight this battle into the last half of the twenty-first century. In light of this, relocation may soon begin to seem like a more reasonable option, we hope. Coastal communities need to include a broader circle of experts in their quest to seek solutions to coastal erosion and global sea level rise.

Get the Corps off the shore. The U.S. Army Corps of Engineers, more or less by default, is the government agency in charge of much of the planning and the funding for the nation's response to sea level rise. It is an agency ill-suited for the job. It has too long a history of checkered competence, high-cost construction, and inefficiency due in significant part to its close dependence on Congress for pork barrel funding, as many critics have pointed out. Part of the problem is that the engineers' "we can fix it" mentality is the wrong mentality for a sensible approach to changing sea level (the agency's motto is "Essayons," a French term meaning "Let us try"). A fundamental reorganization of the Corps is needed.

Take the reins from local government. The problems created by sea level rise are international and national, not local, in scope. Local governments of towns (understandably) follow the self-interests of coastal property owners and developers, so preservation of buildings is inevitably a very high priority. In the debate over which is more important, buildings or beaches, buildings always win. Many tourist beach communities are controlled by a small number of year-round resident voters who make a living from the coastal environment that is utilized by many thousands of nonvoting people during the summer. In Spain, many of the coastal communities are inhabited primarily by nonvoting foreigners. Sea level rise is a national crisis, an issue of national and international interest that must not be solved simply for the benefit of those unwise enough to build at low elevation near an eroding ocean shoreline. In addition, the resources needed to respond to the sea level rise are far beyond those of the local communities.

Sea level rise will threaten coastal ecosystems. Direct destruction of coastal marshes, mangroves, and coral reefs by human activities is currently having a greater impact on these ecosystems than is sea level rise. If we do not immediately act to better protect these critical ecosystems, the combination of rising sea level and human development will destroy them.

Sea level rise provides an opportunity and a challenge for all. Sea level rise does not have to be a natural catastrophe. It could be seen as an opportunity for society to redesign with nature, to anticipate the changes that will occur in the future and to respond in such a fashion as to maintain a coast that future generations will find both useful and enjoyable. It provides a challenge to scientists, planners, environmentalists, politicians, and other citizens alike to stretch the limits of their imagination to respond with flexibility and with careful foresight to development challenges that our society has not faced before. Opportunities will abound for entrepreneurs with fresh ideas on how to live with a rising sea.

The science tells us that the world's shorelines will look different a hundred years from now. These changes need not end the coastal economy as we know it. But preserving our coastal resources and the businesses that depend on them will require insightful and long-term planning. Beginning an honest assessment of how we may deal with inevitable future sea level rise can help ensure that our coastal communities remain the vibrant places that they are today.

References

Chapter 1

Barnett, J., 2001, Adapting to climate change in Pacific Island countries: The problem of uncertainty: *World Development*, v29, p977–93

Connell, J., 2001, An atoll state in peril: *Tiempo*, Issue 42, December—More detailed article in *Journal of Pacific Studies*, 199, v22, p1

Connell, J. & Conway, D., 2000, Migration and remittances in island microstates: *International Journal of Urban Regional Research*, v24, p52–78

Cronin, W.B., 2005, *The disappearing islands of the Chesapeake*: The Johns Hopkins University Press. p 372

Deltawerken online: http://www.deltaworks.org

Dickinson, W.R., 2009, Atoll living: How long already and until when: *GSA Today*, v19, p4–10

Harris, J., 2002, Turning the tide: *Smithsonian Magazine*, v33, p76–88

Kolbert, E., 2006, *Field notes from a catastrophe: Man, nature, and climate change*: Bloomsbury USA, p225

Mason, O.K., 1996, *Geological and anthropological considerations in relocating Shishmaref, Alaska*: Alaska Department of Geological and Geophysical Surveys Report of Investigations 96–7, p18

Pilkey, O.H. & Dixon, K.L., 1996, *The Corps and the shore*: Island Press, p272

Proshutinsky, A., Pavlov, V., & Bourke, R.H., 2001, Sea level rise in the Arctic Ocean: *Geophysical Research Letters*, v28, p2237–40

U.S. Army Corps of Engineers, 2004, *Shishmaref relocation and collocation study*: Alaska District, p75 plus appendices

Chapter 2

Broecker, W.C. & Kunzig, R., 2008, *Fixing climate*: Hill and Wang, p253

Imbrie, J. and Imbrie, K.P., 2005, *Ice ages: Solving the mystery*: Harvard University Press, p224

IPCC, 2007, Climate Change 2007: The Physical Science Basis: Contributions of Working Group I, Fourth Assessment Report: Cambridge University Press, p989

IPCC, 2007, Climate Change 2007: Summary for Policy Makers: Synthesis of the Fourth Assessment Report, http://www.ipcc.ch/pdf/assessment-report/ar4/wg1/ar4-wg1-spm.pdf, p23

Milliman, J.D. & Haq, B.U. (eds), 1996, *Sea-level rise and coastal subsidence: Causes, consequences, and strategies*: Kluwer, p369

Pietrafesa, L.J., Xie, L., & Dickey, D.A., 2005, *On sea level variability on the Eastern Seaboard of the United States*: Conference Proceedings—Solutions to Coastal Disasters 2005, American Society of Civil Engineers, p42–51

Chapter 3

Baker, V., 1994, Geomorphological understanding of floods: *Geomorphology*, v10, p139–56

Bindschalder, R.A. & Bentley, C.R., 2002, On thin ice: *Scientific American*, v287, p98–105

Bruun, P., 1954, Coast erosion and development of beach profiles: *U.S. Army Corps of Engineers, Beach Erosion Board, Technical Memorandum*, v44, p82

Bruun, P., 1962, Sea level rise as a cause of shore erosion: *Proceedings of the American Society of Civil Engineers: Journal of the Waterways and Harbor Division*, v88, p117–30

Chen, J.L., Wilson, C.R., & Tapley, B.D., 2006, Satellite gravity measurements confirm accelerated melting of Greenland ice sheet: *Science*, v314, p1958–60

Cooper, J.A.G. & Pilkey, O.H., 2004, Sea level rise and shoreline retreat: Time to abandon the Bruun Rule: *Global and Planetary Change*, v43, p157–71

Haff, P.K., 1996, Limitations of predictive modeling in geomorphology: In B.L. Rhoads & C.E. Thorn (eds), *The scientific nature of geomorphology*: John Wiley, p337–58

IPCC, 2001, Climate Change 2001: The Physical Science Basis: Contribution of Working Group I, Third Assessment Report, Cambridge University Press, p881

IPCC, 2007, Climate Change 2007: The Physical Science Basis: Contributions of Working Group I, Fourth Assessment Report: Cambridge University Press, p989

IPCC, 2007, Climate Change 2007: Impacts, Adaptation and Vulnerability: Contributions of Working Group II, Fourth Assessment Report: Cambridge University Press, p939

Jasanoff, S., 2007, Technologies of humility: *Nature*, v450, p30

Koutsoyiannis, D., Efstratiadis, A., Mamassis, N., & Christofides, A., 2008, On the credibility of climate predictions: *Hydrological Sciences Journal*, v53, p671–84

Kriegler, E., Hall, J.W., Held, H., Dawson, R., & Schellnhuber, H.J., 2009, Imprecise probability assessment of tipping points in the climate system: *Proceedings of the National Academy of Sciences* [doi: 10.1073/pnas.0809117106]

Liu, J., et al., 2007, Complexity of coupled human and natural systems: *Science*, v317, p1513–16

Oreskes, N.K., 1998, Evaluation (not validation) of quantitative models: *Environmental Health Perspectives*, v106, p1453–60

Pearce, Fred, 2008, Are climate scientists overstating their models? An interview with physicist Lenny Smith: *New Scientist Magazine*, Issue 2685, p42–43

Pilkey, O.H. & Pilkey-Jarvis, L., 2007, *Useless arithmetic: Why environmental scientists can't predict the future*: Columbia University Press, p230

Rahmstorf, S., et al., 2007, Recent climate observations compared to projections: *Science*, v316, p709

Rahmstorf, S., 2007, A semi-empirical approach to projecting future sea level rise: *Science*, v315, p368–70

Sayes, S. & Pye, K., 2007, Implications of sea level rise from coastal dune habitat conservation in Wales, UK: *Journal of Coastal Conservation*, vii, p31–52

Science and Technology Committee, 2008, *Statement on Sea Level Rise in the Coming Century Report*: Miami Dade County Climate Change Task Force, January 17, 2008, p9

Chapter 4

Chen, J.L., Wilson, C.R., & Tapley, B.D., 2006, Satellite gravity measurements confirm accelerated melting of Greenland ice sheet: *Science*, v313, no.5795, p1958–60

Davis, C., et al., 2005, Snowfall-driven growth in East Antarctic ice sheet mitigates recent sea level rise: *Science*, v308, p1898–1901

Dyurgerov, M.B. & Meier, M.F. (2000), Twentieth-century climate change: Evidence from small glaciers: *Proceedings of the National Academy of Sciences*, v97, p1406–11

Goldstein, R.M., Englehardt, H., Barclay, K., & Frolich, R.M., 1993, Satellite radar interferometry for monitoring ice sheet motion: Application to an Antarctic ice stream: *Science*, v262, p1525–30

Grinsted, A., et al., 2009, Sea level rise of one meter within 100 years: *Science Daily*, January 12, 2009

Joughin, I., Abdalati, W., & Fahnestock, M., 2004, Large fluctuations in speed on Greenland's Jacobshavn Isbrae glacier: *Nature*, v432, p608–10

Krabill, W., et al., 1999, Rapid thinning of parts of the southern Greenland ice sheet: *Science*, v283, p1522–24

Molnia, B., 2008, Glacier and landscape change in response to changing climate: U.S. Geological Survey, www.usgs.gov/global_change/glaciers

Molnia, B., 2008, Glaciers of Alaska: In R.S. Williams & J.G. Ferrigno (eds), Satellite image atlas of glaciers of the world: *U.S. Geological Survey Professional Paper 1386K*, 525p

Oppenheimer, M., 1998, Global warming and the stability of the West Antarctic ice sheet: *Nature*, v393, 28 May, p325–32

Rignot, Eric, et al., 2004, Accelerated ice discharge from the Antarctic Peninsula following the collapse of Larsen B Ice Shelf: *Geophysical Research Letters*, v31, L18401 [doi:10.1029/2004GL020697]

Rignot, E. & Kanagaratnam, P., 2006, Changes in the velocity structure of the Greenland ice sheet: *Science*, v311, p986–90 [doi: 10.1126/science.1121381]

Rignot, E., et al., 2008, Recent Antarctic ice mass loss from radar interferometry and regional climate modelling: *Nature Geoscience*, v1, 106-10 [doi:10.1038/ngeo102]

Shepherd, A. & Wingham, D., 2007, Recent sea level contributions of the Antarctic and Greenland ice sheets: *Science*, v315, p1529–32

Vaughan, D.G., 2005, How does the Antarctic ice sheet affect sea level rise? *Science*, v308, 24 June, p1877–1878

Vaughan, D.G. & Arthern, R., 2007, Why is it hard to predict the future of ice sheets? *Science*, v315, 1503–4 [doi:10.1126/science.1141111]

Vaughn, D., Holt, J.W., & Blankenship, D.D., 2007, West Antarctic links to sea level estimation: *EOS, Transactions of the American Geophysical Union*, v88, 13 Nov 2007.

Velicogna, I. & Wahr, J., 2006, Measurements of time-variable gravity show mass loss in Antarctica: *Science*, v311, no.5768, p1754–56

Chapter 5

Armstrong, J.S. (ed), 2001, *Principles of forecasting: A handbook for researchers and practitioners*: Springer, p849

Armstrong, J.S., Green, K.C., & Soon, W., 2008, Polar bear population forecasts: A public policy forecasting audit: Interfacer, v38, p382–404

Bowen, M., 2008, *Censoring science*: Dutton, p324

Ceccarelli, L., 2001, *The shaping of science with rhetoric: The cases of Dobzhansky, Schrodinger, and Wilson*: University of Chicago Press, p192

Beck, G., 2007, *An inconvenient book: Real solutions to the world's biggest problems*: Simon and Schuster, p295

Gore, A., 2006, *An inconvenient truth: The planetary emergency of global warming and what we can do*: Rodale Books, p328

Hansen, J.E., 2007, Scientific reticence and sea level rise: *New Scientist Magazine*, v2614, p30–34

Horner, C., 2007, *The politically incorrect guide to global warming and environmentalism*: Regnery Publishers, p366

Lomborg, B., 2001, *The skeptical environmentalist: Measuring the real state of the world*: Cambridge University Press, p540

Lomborg, B., 2007, *Cool it: The skeptical environmentalist's guide to global warming*: Knopf, p272

Michaels, D., 2005, Doubt is their product: *Scientific American*, v292, p96–101

Michaels, P., 2004, *Meltdown: The predictable distortion of global warming by scientists, politicians, and the media*: The Cato Institute, p208

Michaels, P., 2005, *Shattered consensus: The true state of global warming*: Rowman and Littlefield, p304

Monckton, C., 2007, *35 inconvenient truths: The errors in Al Gore's movie*: The Science and Public Policy Institute, p21

Nordhaus, W.D., 2008, *A question of balance: Weighing the options on global warming policies*: Yale University Press, p256

Oreskes, N., 2004, Science and public policy: What's proof got to do with it? *Environmental Science & Policy*, v7, p369–83

Rahmstorf, S., 2007, A semi-empirical approach to projecting future sea level rise: *Science*, v315, p368–70

Robinson, A.B., Robinson, N.E., & Soon, W., 2007, Environmental effects of increased atmospheric carbon dioxide: *Journal of American Physicians and Surgeons*, v12, p79–90

Romm, J., 2007, *Hell and high water—and what you should do*: HarperCollins Publishers, p285

Singer, S.F. & Avery, D., 2008, *Unstoppable global warming: Every 1,500 years*: Rowman and Littlefield Publishers, p264

Solomon, L., 2008, *The deniers: The world renowned scientists who stood *against global warming, hysteria, political persecution, and fraud*: Richard Vigilante Books, p239

Union of Concerned Scientists, 2007, *Smoke, mirrors, and hot air*: www. ucsusa.org, p63

Chapter 6

Alleng, G.P., 1998, Historical development of the Port Royal mangrove wetland, Jamaica: *Journal of Coastal Research*, v14, p951–59

Bacon, P.R., 1994, Template for evaluation of impacts of sea level rise on Caribbean coastal wetlands: *Ecological Engineering*, v3, p171–86

Blasco, F., Saenger, P., & Janodet, E., 1996, Mangroves as indicators of coastal change: *Catena*, v27, p167–78

Carson, R., 1955, *The edge of the sea*: Houghton Mifflin, p276

Ellison, A.M. & Farnsworth, E.J., 1996, Anthropogenic disturbance of Caribbean mangrove systems: Past impacts, present trends, and future predictions: *Biotropica*, v28, p549–65

Ellison, J.C. & Stoddart, D.R., 1991, Mangrove ecosystem collapse during predicted sea level rise? Holocene analogs and implications: *Journal of Coastal Research*, v7, p151–65

Field, C.D., 1995, Impact of expected climate change on mangroves: *Hydrobiologia*, v295, p75–81

Jin-Eong, O., 2004, The ecology of mangrove conservation and management: *Hydrobiologia*, v295, p343–51

Kolbert, E., 2006, The darkening sea: What carbon emissions are doing to the sea: *The New Yorker*, November 20, 2006, p66–75

Macintyre, I.G., 2007, Demise, regeneration and survival of some Western Atlantic reefs during the Holocene transgression: In R.B. Aronson (ed), *Geological approaches to coral reef ecology*, Ecological Studies 192: Springer, p181–200

Parkinson, R.D., Delaune, R.D., & White, J.R., 1994, Holocene sea level rise and the fate of mangrove forests within the wider Caribbean region: *Journal of Coastal Research*, v10, p1077–1086

Pandolfi, J.M., et al., 2006, Mass mortality following disturbance in Holocene coral reefs from Papua, New Guinea: *Geology*, v34, p949–52

Semenuik, V., 1994, Predicting the effect of sea level rise on mangroves in Northwestern Australia: *Journal of Coastal Research*, v10, p1050–1076

Snedaker, S.C., 1995, Mangrove and climate change in the Florida and Caribbean region: Scenario and hypotheses discussion: *Hydrobiologia*, v295, p43–49

Spencer, C., 2008, *Edisto Island, 1663 to 1860: Wild Eden to cotton aristocracy*: The History Press, p222

Titus, J.G. (ed), 1988, *Greenhouse effect sea level rise and coastal wetlands*: U.S. Environmental Protection Agency, p152

Wood, C.M. & McDonald, D.G. (eds), 1997, *Global warming: The implications for fresh water and marine fish*: Cambridge University Press, p425

Woodroffe, C.D., 1990, The impact of sea level rise on mangrove shorelines: *Progress in Physical Geography*, v14, p483–520

Chapter 7

Cline, W.R., 2007, *Global warming and agriculture: Impact estimated by country*: Peterson Institute, p201

Dasguopta, S., Laplante, B., Meissner, C., Wheeler, D., & Yan, J., 2007, The impact of sea level on developing countries: A comparative analysis: World Bank Policy Research Working Paper 4136, p51

Emanuel, K.A., 1987, The dependence of hurricane intensity on climate: *Nature*, v326, p483–85

Francis, J.A. & Hunter, E., 2006, New insight into the disappearing Arctic Sea ice: *EOS, Transactions of the American Geophysical Union*, v87, p509–11

Gibbons, S.J.A. & Nicholls, R.J., 2006, Island abandonment and sea level rise: An historical analog from the Chesapeake Bay, USA: *Global Environmental Change*, v16, p40–47

Jacob, K., Gornitz, V., & Rosenzweig, C., 2007, Vulnerability of the New York City metropolitan area to coastal hazards, including sea-level rise: Inferences for urban coastal risk management and adaptation policies: In L. McFadden, R. Nicholls, & E. Penning-Rowsell (eds), *Managing coastal vulnerability*: Elsevier, p139–56

Kerr, R.A., 2006, Global warming may be homing in on Atlantic hurricanes: *Science*, v314, p910–11

Kister, C., 2004, *Arctic melting: How global warming is destroying one of the world's largest areas*: Common Courage Press, p224

Komar, P., 1997, *The Pacific Northwest coast*: Duke University Press, p195

Milliman, J.D. & Haq, B.U. (eds), 1996, *Sea level rise and coastal subsidence: Causes, consequences and strategies*: Kluwer Academic Publishers, p369

Nicholls, R.J., et al., 2007, Ranking of the world's cities most exposed to coastal flooding today and in the future: OECD, Environment Working Paper No. 1, Executive Summary, p12

Nishioka, S. & Harasawa, H. (eds), 2000, Global warming: The potential impact on Japan: *Climatic Change*, v47, p213–15

Pilkey, O.H. & Dixon, K.L., 1996, *The Corps and the Shore*: Island Press, p272

Riggs, S.R., Cleary, W.J., & Snyder, S.W. 1995, Influence of inherited geologic framework upon barrier beach morphology and shoreface dynamics: *Marine Geology*, v126, p213–34

Sarwar, G.M., 2005, *Impacts of sea level rise on the coastal zone of Bangladesh*: Masters Thesis, Lund University, Sweden, p38

Wind, H.G. (ed), 1987, *Impact of sea level rise on society*: A.A. Balkema Publishers, p191

Chapter 8

Coastal Protection and Restoration Authority of Louisiana, 2007, *Integrated ecosystem restoration and hurricane protection: Louisiana's Comprehensive Master Plan for a Sustainable Coast*: Coastal Protection and Restoration Authority of Louisiana

Lewis, P.F., 2003, *New Orleans: The making of an urban landscape*: Center for American Places, p200.

Louisiana Coastal Wetlands Conservation and Restoration Task Force, 1998, *Coast 2050: Toward a sustainable coastal Louisiana*: Louisiana Department of Natural Resources, p161

Louisiana Wetland Protection Panel, 1987, Saving Louisiana's coastal wetlands: U.S. Environmental Protection Agency (EPA-230-02-87-026), p102

McNabb, D. & Madere, L.E., 2003, A history of New Orleans: Accessed at www.madere.com/history.html, January 25, 2009

Morgan, J.P. & Morgan, D.J., 1983, *Accelerating retreat rates along Louisiana's coast*: Louisiana Sea Grant College Program, p41

Morton, R.A., Tiling, G., & Ferina, N.F., 2003, Primary causes of wetland loss at Madison Bay, Terrebonne Parish, Louisiana: United States Geological Survey Open File Report 03-60, p48

National Research Council, 2006, *Drawing Louisiana's new map: Addressing land loss in coastal Louisiana*: National Academies Press, p204

National Research Council, 2008, *First Report from the NRC Committee on the Review of the Louisiana Coastal Protection and Restoration (LACPR) Program*: National Academies Press, p32

Penland, S., Ramsey, K.E., McBride, R.A., Moslow, T.F., & Westphal, K.A., 1989, *Relative sea level rise and subsidence in Louisiana and the Gulf of Mexico*: Louisiana Geological Survey, p65

Program for the Study of Developed Shorelines, 2008, Expert panel assessment of coastal hazard mapping in the southeastern U.S.: Results from the *First Cullowhee Coastal Conference July 23–24, 2007*: Available at psds. wcu.edu, p10

U.S. Army Corps of Engineers, 1963, Hurricane study: Interlying area along coastal Louisiana in the vicinity of Houma, U.S. Army Corps of Engineers District, New Orleans File No. H-2-2283

Wilkins, J.G., Emmer, R.E., Hwang, D.J., Kemp, G.P., Kennedy, B., Hassan, M., & Sharky, B., 2008, *Louisiana coastal hazard mitigation guidebook*: Louisiana Sea Grant College Program, p246

Chapter 9

Barkham, P., 2008, Waves of destruction, *The Guardian*, 17 April, www. guardian.co.uk/environment/2008/apr/17/flooding.climate

Burian, S.J., 2000, Urban wastewater management in the United States: Past, present and future: *Journal of Urban Technology*, v7, p33–62

Bush, D.M., Pilkey, O.H., & Neal, W.J., 1996, *Living by the rules of the sea*: Duke University Press, p179

Clover, Charles, "Norfolk Broads Could Be Lost to Sea in a Year," *Daily Telegraph*, March 31, 2008. http://www.telegraph.co.uk/earth/earth-news/3337776/Norfolk-Broads-Could-Be-Lost-to-Sea-in-a-year. html

Dean, C., 1999, *Against the tide: The battle for America's beaches*: Columbia University Press, p234

Girling, R., 2007, *Sea change: Britain's coastal catastrophe*: Transworld Publishers, p353

Howard, J.D., Kaufman, W., & Pilkey, O.H., 1985, A national strategy for beach preservation White Paper: Second Skidaway Institute of Oceanography Conference on America's Eroding Shoreline

Houck, O., 2006, Can we save New Orleans? *Tulane Environmental Law Journal*, v19, p68

IPCC, 2005, *Special report on carbon dioxide capture and storage*: Cambridge University Press, p429

Kaufman, W. & Pilkey, O.H., 1979, *The beaches are moving: The drowning of America's shoreline*: Duke University Press, p336

Keith, D.W., 2000, Geoengineering the climate: History and prospect: *Annual Review of Energy and the Environment*, p245–84

National Research Council, Marine Board, Committee on Engineering Implications, 1987, *Responding to changes in sea level: Engineering implications*: National Academy of Sciences

Program for the Study of Developed Shorelines: Beach Nourishment Table, http://psds.wcu.edu

Romm, J., 2008, *Hell and high water: The global warming solution*: Harper Perennial, p304

Sargent, W., 2007, *Just seconds from the ocean: Coastal living in the wake of Katrina*: University Press of New England, p142

Schmidt, G., 2008, Hypothesis testing and long range memory: *Real Climate*, http://www.realclimate.org

Sengupta, B., 2009, In silt, Bangladesh sees potential shield against climate shift, *New York Times*, March 20, p.A16

Woodwell, G.M., 1991, Forests in a warming world: A time for new policies: *Climatic Change*, v19, p245-51

Keywords

Sea level rise

West Antarctic ice sheet

Greenland ice sheet

Shoreline erosion

Flooding

Barrier islands

Global warming

Climate change

Climate skeptics

Coastal hazards

Coastal management

Acknowledgments

We owe much to many. Owen Mason educated us about and led us to shorelines of Arctic Alaska. Tony Weyiouanna, Inupiat Eskimo and able spokesperson for the village of Shishmaref, Alaska, explained the dilemma of coastal villages in a time of sea level rise and melting permafrost. Bruce Molnia's latest treatise on Alaskan glaciers may be the heaviest (literally) book we've seen in years. His work on mountain glacier melting was important to us, and we gained much from his editorial comments on an early version of our manuscript. Abby Sallenger and Robert Morton of the U.S. Geological Survey, Denise Read of the University of New Orleans, and Angelina Freeman of the Environmental Defense Fund provided much insight on the Mississippi Delta and Gulf Coast sea-level-rise problems. Hal Wanless, of the University of Miami and a voice of reason in the wilderness of Florida politics, inspired us with his efforts to educate the public about the inevitability of future sea level rise. Ian Macintyre, Smithsonian geologist, still diving on coral reefs in his midseventies, was our coral reef adviser. Wallace Kaufman kept us abreast of the critical view of global change with a flood of articles from those who manufacture doubt and from those who really did doubt the science. Dame Jane Resture, South Pacific Islander, allowed us to use her beautiful poem about Tuvalu. Robert Thieler, Peter Haff, Thomas Crowley, Joseph Kelley, Andrew Cooper, David Bush, and others too numerous to list advised on a variety of issues along the way.

Jonathan Cobb at Island Press saw the promise in this book during a phone conversation a couple of years back and has guided and encouraged us ever since. A good editor has to take pains to keep the authors honest and not let their excitement for the topic get out of hand. Jonathan did so with great diplomacy and became a friend in the process. Sharlene Pilkey spent hours Googling after sometimes very obscure facts. Norma Longo also provided extensive invaluable background research and was a very attentive copy editor. Holli Thompson, Norma Longo, Bruce Molnia, Ian Macintyre, Sharlene Pilkey, Keith Pilkey, Len Pietrafesa, and Leigh Anne Young provided editing on the early chapter drafts. We both owe a great deal to the support and assistance of our spouses, Leigh Anne Young and Sharlene Pilkey.

Index

About Island Press

Since 1984, the nonprofit Island Press has been stimulating, shaping, and communicating the ideas that are essential for solving environmental problems worldwide. With more than 800 titles in print and some 40 new releases each year, we are the nation's leading publisher on environmental issues. We identify innovative thinkers and emerging trends in the environmental field. We work with world-renowned experts and authors to develop cross-disciplinary solutions to environmental challenges.

Island Press designs and implements coordinated book publication campaigns in order to communicate our critical messages in print, in person, and online using the latest technologies, programs, and the media. Our goal: to reach targeted audiences—scientists, policymakers, environmental advocates, the media, and concerned citizens—who can and will take action to protect the plants and animals that enrich our world, the ecosystems we need to survive, the water we drink, and the air we breathe.

Island Press gratefully acknowledges the support of its work by the Agua Fund, Inc., Annenberg Foundation, The Christensen Fund, The Nathan Cummings Foundation, The Geraldine R. Dodge Foundation, Doris Duke Charitable Foundation, The Educational Foundation of America, Betsy and Jesse Fink Foundation, The William and Flora Hewlett Foundation, The Kendeda Fund, The Andrew W. Mellon Foundation, The Curtis and Edith Munson Foundation, Oak Foundation, The Overbrook Foundation, the David and Lucile Packard Foundation, The Summit Fund of Washington, Trust for Architectural Easements, Wallace Global Fund, The Winslow Foundation, and other generous donors.

The opinions expressed in this book are those of the author(s) and do not necessarily reflect the views of our donors.